Rufino C. Pabico, M. D.
University of Rochester
Medical Center
Rochester, N. Y. 14620

Animal Models for Biomedical Research III

PROCEEDINGS OF A SYMPOSIUM

Sponsored jointly by the

INSTITUTE OF LABORATORY ANIMAL RESOURCES
National Research Council

and the

AMERICAN COLLEGE OF LABORATORY ANIMAL MEDICINE

NATIONAL ACADEMY OF SCIENCES
Washington, D.C.
1970

This report was supported in part by Contract PH43-64-44 with Chemotherapy, National Cancer Institute, and the Animal Resources Branch, National Institutes of Health, United States Public Health Service; Grant #RC-1L from the American Cancer Society, Inc.; Contract Nonr-2300(24) with the Office of Naval Research, the United States Army Medical Research and Development Command, and the United States Air Force; Contract AT(49-7)643 with the United States Atomic Energy Commission; Contract 12-14-110-3726-91 with the United States Department of Agriculture; and Contract NSF-C310, Task Order No. 173, with the National Science Foundation.

International Standard Book Number 0-309-01854-4

Available from
Printing and Publishing Office
National Academy of Sciences
2101 Constitution Avenue
Washington, D.C.   20418

Library of Congress Card Number 76-607190

Preface

This symposium, the third entitled "Animal Models for Biomedical Research," was held July 14, 1969, in Minneapolis, Minnesota, in conjunction with the annual meeting of the American Veterinary Medical Association. It was sponsored by the American College of Laboratory Animal Medicine and the Institute of Laboratory Animal Resources (ILAR) of the National Research Council.

The program was designed to emphasize the fact that unique animal models exist for a very diverse biomedical need. It is hoped that this symposium will stimulate investigators to utilize existing models where they apply and to seek out new models for studies that as yet do not have good animal models.

Sincere appreciation is expressed to the authors for their efforts and to Dr. Robert H. Yager, Executive Secretary of ILAR, for his help throughout.

CHARLES C. MIDDLETON, D.V.M.
Symposium Chairman

Contents

The Chediak-Higashi Syndrome: A Review of the Disease
in Man, Mink, Cattle, and Mice   1
    George A. Padgett, James M. Holland, David J. Prieur,
    William C. Davis, John R. Gorham

Hereditary Hyperbilirubinemia in Sheep   13
    Charles E. Cornelius

Animal Models of Atherosclerosis   22
    Thomas B. Clarkson, Robert W. Prichard, Bill C. Bullock,
    Noel D. M. Lehner, Hugh B. Lofland, Richard W. St. Clair

Animal Models in Hemophilia Research: A Review   42
    Merle E. Muhrer

Animal Models for the Study of Hemoglobinopathies   52
    Hyram Kitchen, Ethel Oliver Kouba, Caroline W. Easley

Cyclic Neutropenia in Man and Dog   71
    John E. Lund

The Ehlers-Danlos Syndrome of Dogs and Mink   80
    Gerald A. Hegreberg, George A. Padgett, Roy C. Page

Psychological Factors in Comparative Biomedical Research   91
    Robert Ader

The Use of Amphibians in Biomedical Research   103
  George W. Nace

Animal Models for Pharmacotherapeutic Studies   125
  David P. Rall

The Use of Marine Mammals in Biomedical Research   133
  Richard C. Hubbard

Symposium Contributors   147

The Chediak-Higashi Syndrome:
A Review of the Disease
in Man, Mink, Cattle, and Mice

GEORGE A. PADGETT, JAMES M. HOLLAND,
DAVID J. PRIEUR, WILLIAM C. DAVIS, and
JOHN R. GORHAM

The Chediak-Higashi syndrome (C-HS) is an intriguing genetic disease that has been reported to occur in four species: man,[1] mink,[2] cattle,[3] and mice.[4] The disease is inherited as an autosomal recessive trait in all four species and is characterized by abnormally large intracytoplasmic granules in nearly all cell types that normally form such structures. In addition, affected individuals are partial albinos and are more susceptible than are normal ones to a variety of infectious agents. The disease has been reviewed recently[5]; therefore, the general aspects of this syndrome will not be considered here. The purpose of this article is to discuss the value of studying this disease in animals, not only as it relates to the C-HS in man, but also to point out its usefulness as a model to evaluate developmental and functional processes in various cell types. The discussion will cover four points: Homology of the syndrome among the four affected species, the mechanism of lysosomal formation, the mechanism of lysosomal function, and potential uses for animals with the C-HS.

HOMOLOGY OF THE C-HS

The amount of information available for the four species varies considerably. The most extensive clinical studies have been made in man, next most in mink and cattle. Little clinical information, including that associated with increased susceptibility to disease, is available on the mouse. With the exception of a generalized reticulohistiocytic infiltrate, which has been reported to occur only in man, the general clinical findings have been remarkably similar in man, mink, and cattle.[6] Experimental studies have been more extensive in mink and cattle than in man. Where correlations among the three species are possible, the results are generally in agreement. The major differences that have been reported could usually be related to secondary diseases or to chemotherapeutic agents used to treat the diseases in affected members of all three species. In addition, there is the problem of species specificity of an agent; for example, Aleutian disease does not occur in cattle or man, only in mink, and the generalized infiltrate in man referred to earlier could easily be due to an agent that does not affect mink, cattle, or mice. This does not mean that the C-HS is different in the various species in which it has been observed; it does suggest, however, that, although four species have a similar genetic trait, this fact does not necessarily alter their responses to an agent that normally affects only one of them.

The last category in which comparisons between the species are clearly possible involves electron microscopy. The most extensive studies to date are those of Lutzner[7] on mink and Doak[8] on cattle. These studies are in agreement as to the distribution of abnormal intracytoplasmic organelles in the various cell types in affected members of both populations. Studies in man have been limited to the periph-

TABLE 1   Similarities Suggesting that the Chediak-Higashi Syndrome Is Homologous in Man, Mink, Cattle, and Mice

Characteristic	Man	Mink	Cattle	Mice
Mode of inheritance (autosomal recessive)	yes	yes	yes	yes
Susceptible to various infectious diseases	yes	yes	yes	—
Bactericidal ability of neutrophils normal	yes	yes	yes	—
Chemotaxis normal	yes	yes	—	—
Normal immunologic response	yes	yes	yes	—
Pigmentary dilution	yes	yes	yes	yes
Bleeding tendency	yes	yes	yes	—
Abnormal granules found in the same cell types	yes	yes	yes	yes
Morphology of abnormal granules similar	yes	yes	yes	yes
Histochemistry of abnormal granules similar between species and the same as found in normal granules	yes	yes	yes	yes

— Unknown or not reported.

TABLE 2   Differences that Might Suggest that the Chediak-Higashi
Syndrome Is Not Homologous in Man, Mink, Cattle, and Mice

Characteristic	Man	Mink	Cattle	Mice
Reticulohistiocytic infiltrate (also called malignant lymphoma by some investigators)	yes	no	no	no
Aleutian disease of mink	no	yes	no	no
Virus-like particles observed by electron microscopy	yes[1]	no	yes	no
Sequestration vacuoles in the cytoplasm of leukocytes	yes[1]	no	no	no
Degeneration of the abnormal granules	yes[1]	no	no	no
Increased lysozyme levels	yes	no	no	no
Abnormal serum lipid pattern	yes	no	no	no
Autophagy	yes[1]	no	no	—
Abnormally permeable or fragile granule membrane	yes[1]	no	no	—

1. Reported by one group of investigators; see References 15, 16, 17, 45.
— Unknown or not reported.

eral blood and bone marrow and there is considerable disagreement—
not so much as to what was observed, but more particularly the interpretation of the observations. The electron microscopic observations
are discussed more extensively in the following sections. Similarities
and differences that have been reported to occur in the C-HS are listed
in Tables 1 and 2.

GRANULOGENESIS

Intracytoplasmic secretory granules comprise a large and diverse group
of subcellular organelles common to many mammalian cells. While the
ultimate fate of secretory granules varies with cell type, and in the
same cell at different times, it is generally believed the processes involved in their synthesis are related.[9] Ultrastructural autoradiographic
and cytochemical studies provide the following temporal and spatial
summary of granule formation: Protein, synthesized by polyribosomal
clusters lining the endoplasmic reticulum, moves through the channels
of the reticular cisternae to the Golgi complex where it is concentrated, wrapped in a unit membrane package, and released as a finished

granule.[10] Little is known about the genetic control of granule formation and function. We suggest that the C-HS provides a unique opportunity to gain valuable insight into these fundamental life processes.

Anomalous inclusions in white blood cells were instrumental in establishing the C-HS as a disease entity in both man and animals.[1, 3, 4] Except for size, these inclusions resembled the normal lysosomal granules of the neutrophilic leukocyte.[11, 12] Subsequent studies on human and animal cases of the syndrome have emphasized the presence of anomalous granules in a variety of cells.[7, 8, 13] Various hypotheses have been advanced to explain the occurrence of bizarre granules in many different cell types. There is little probability that humoral or hormonal factors are involved, since inclusions have been observed to persist in cells obtained from both humans and animals when the C-HS is maintained in tissue culture for long periods.[14, 15] An increase in the latent phagocytic propensity of the cell, with an accumulation of phagolysosomes or residual bodies, has been proposed, as has been a decrease in the stability of granules and leakage of the granule content into the cytoplasm, promoting the formation of autophagic vacuoles.[16, 17] Neither of these possibilities is likely, however, for the simple reason that inclusions are commonly found in cells that either are not phagocytic or that contain granules whose content is normally inert in the cytoplasm. Rarely have C-HS inclusions been demonstrated to contain recognizable cell structures, an essential condition if one is to evoke an autophagocytic hypothesis. Evidence that the inclusions represent an accumulating metabolic byproduct are also unconvincing.[18, 19] A consistent and serious flaw in all of these hypotheses is that they overlook the occurrence of inclusions in diverse cell types, and, further, that the inclusions often resemble those native to the cell in question. It would seem that any valid working hypothesis should consider these facts. In formulating such a hypothesis, one is forced to look to something that all involved cells have in common, namely, a Golgi-dependent, granule-synthesizing pathway.

How might the Golgi complex be involved in the C-HS? As suggested by Lutzner *et al.*[7]: "A defect in the transfer of material from endoplasmic reticulum to Golgi systems, a defect in the process of granule assembly, or a defect in granule membrane formation could conceivably result from the presence of a mutant gene, could be common to many cell types, and might explain all the bizarre patterns observed." This brings us to the point of testing any hypothesis of abnormal membrane fusion or synthesis.

The rabbit heterophile is the functional analogue of the neutrophilic leukocyte found in other mammals. Autoradiographic and cytochemical studies have confirmed that the Golgi complex plays a central role in granule synthesis.[20] Evidence that the granule complement of the cell is heterogenous is provided by ultrastructural histochemistry of mature cells and their progenitors in bone marrow preparations.[21, 22, 23] Ultrastructural cytochemical properties of at least three distinct granule classes are defined in a recent review on this subject.[24] Normal rabbit heterophile granules vary enough in size, shape, and density that they may be resolved into distinct populations by continuous flow zonal centrifugation on sucrose gradients.[25] Ultracentrifugal studies agree well with ultrastructural observations. Granules thus resolved vary not only in enzyme content, but also in time of formation. Primary granules are rich in acid hydrolases characteristic of the typical lysosome.[26] These granules are large when compared to other granules in the cell, and are formed only during the promyelocyte stage by a process, first, of budding from the proximal or convex face of the Golgi complex, and then eventual aggregation of vacuoles with a dense core. Secondary granules are smaller, more numerous, and contain alkaline phosphatase. They are formed during the myelocyte and subsequent stages of cell maturation from the distal or convex face of the Golgi complex by budding and by the confluence of small granules of moderate density.[27] A small, moderately dense, acid phosphatase reactive granule occurs in small numbers in mature heterophils.[28]

These observations offer an intriguing explanation for the morphologic appearance of the C-HS leukocyte. Only a few very large granules are observed against a background of numerous smaller granules. Preliminary ultrastructural studies on C-HS mink bone-marrow preparations, paralleling the work done on the rabbit heterophil, indicate that the defect in granule synthesis is limited to the primary granule in this cell.[29] The fact that only the primary granules are affected may provide insight into the genetic control of granule synthesis. Are primary granules synthesized normally, and do they then undergo unregulated fusion to reach the large size attained in mature cells? A reasonable speculation would predict that control of membrane fusion and stability is an intrinsic property of the membrane proper. If a mutant gene can effectively alter Golgi-mediated membrane synthesis, the characterization of this alteration in C-HS animals will be of definite assistance in defining the process of membrane synthesis in normal individuals.

CELL FUNCTION

Given the fact that formation of abnormal granules in the C-HS is due to an aberration of the normal granule forming mechanism (specifically, a failure to control the continued fusion of vesicles, resulting in massive azurophil granules), the relationship of this malfunction to increased disease susceptibility is still obscure.

Several studies suggest a relationship between the abnormally large granules and disease states but at present the evidence directly linking them is tenuous or unconfirmed.

In support of a previously mentioned hypothesis, proposed to explain defective leukocyte function in children with the C-HS, electron microscopic evidence has been presented that suggests that enlarged lysosomes degenerate *in vivo*, and their content leaks through an abnormally permeable membrane. This leakage of hydrolytic enzymes damages cytoplasmic constituents, thereby altering all functions.[17] A similar process was suggested to relate a degenerative change in Schwann cells, in a child with the C-HS, to the development of a peripheral neuropathy in that child.[30] However, subsequent examination of autopsy material from this and another case led to the interpretation that the neuropathy was due instead to a diffuse lymphohistiocytic infiltrate of the nerve, i.e., a manifestation of secondary disease.[31, 32]

Increased serum lysozyme levels in four children with the C-HS has been reported.[33] The increase has been interpreted to reflect an excessive granulocyte turnover due to increased intramedullary granulocyte destruction or autophagy leading ultimately to cell damage and lysis. No differences in serum lysozyme levels were reported in C-HS mice, mink, or cattle.[12, 34] These findings may indicate a basic difference in the syndrome in man as compared with animals. Alternative explanations are that this difference reflects the presence of a secondary disease in the children, or that the increased lysozyme results from chemotherapeutic agents used to treat the secondary disease. It is of interest in this regard that the two patients with a frank accelerated or lymphomatous phase had significantly higher levels of lysozyme than the other two patients, and that elevated lysozyme levels do occur in several of the leukemias. Secondary diseases were not present in the mice, mink, and cattle studied in regard to lysozyme levels, nor had they been treated with cortisone or antitumor agents.

An increased release of bound enzyme from liver lysosomes, mitochondria, and peroxisomes was found in C-HS mink when compared with normal mink.[35] It was concluded that the more rapid release of enzymes from C-HS liver lysosomes may be because an abnormally permeable membrane surrounds these structures, and that the involvement of mitochondria and peroxisomes may be secondary to autophagy caused by the *in vivo* release of lysosomal enzymes.

There are at present insufficient data either to implicate strongly or to rule out the suggestion that autophagy is a major functional aberration in the C-HS. If such is the case, however, the C-HS provides a unique model for the study of a basic process that in normal cells provides a mechanism that protects the cell from degenerate intracytoplasmic organelles.

There is good clinical evidence for increased susceptibility to bacterial organisms in children, mink, and cattle with the C-HS. Within the limitations of the test system employed, however, no *in vitro* difference was reported when the bactericidal capacity of neutrophils obtained from C-HS children, mink, and cattle was compared with normal controls.[12, 36] Following the uptake of particulate material by normal neutrophilic leukocytes, the cells normally degranulate by a process of granule fusion with the membrane enveloping the foreign material. Degranulation is impaired in C-HS mink, cattle, and children.[12, 37] Normal-sized granules in C-HS neutrophils were observed to fuse with the phagocytic vacuole, while the abnormally large granules in the same cell either would not lyse or lysed at a slower rate than normal.

It is of interest that the cationic bactericidal proteins of the rabbit heterophil leukocyte appear to be localized in the specific granule of these cells.[25, 38] If this is also true of mink, cattle, and man, it provides an explanation for the failure to find a defect in bactericidal capacity of C-HS neutrophils since in C-HS individuals specific granules are apparently normal both in structure and function.

It may well be that the major problem in the C-HS is not related to the inability to kill microorganisms, but that this defect reflects the inability of abnormal lysosomes to catabolize secondary products associated either with the organism or with the host response to the organism. Aleutian disease of mink presents a situation in which the host response apparently overloads a catabolic process in the mesangial cell. A more rapid progression of lesions and a shorter mean death time is seen in C-HS mink when compared with non-C-HS mink.[39] Available evidence suggests that death in both C-HS and non-C-HS mink probably is related to the glomerular lesions.[40, 41] Furthermore, it has been shown that the glomerular lesions result from the deposition of macromolecular complexes in the mesangium.[41] These complexes are composed of complement, gamma globulin, and probably the agent of the disease. The rate of replication and the amount of circulating agent in the peripheral blood is similar when-C-HS and non-C-HS mink are compared.[42] If the stoichiometric relationship of the agent, gamma globulin, and complement is similar in C-HS and non-C-HS mink, the fact that there are similar amounts of the agent circulating in the peripheral blood suggests that the defect involves impaired mesangial clearance of the complexes, and not an increased rate of deposition. Recent work supports the concept of ineffectual catabolism due to a lysosomal defect. When intraperitoneally inoculated egg albumen was taken up by the proximal renal tubular epithelium of C-HS and non-C-HS mink and mice, clearance was delayed in C-HS mink and mice when compared to non-C-HS animals of both species.[43] If this delay in catabolism results from a delay, or failure, of lysosomes to fuse with pinocytosed protein, thus withholding the essential degradative enzymes, it would tend to substantiate the contention put forth for the difference in susceptibility of C-HS and non-C-HS mink to Aleutian disease. However, considerably more work is required before definitive statements can be made in this regard.

Several investigators have postulated a membrane defect as the primary abnormality in the C-HS.[12, 13, 16, 17, 43] None of the work is sufficiently definitive, however, to prove or disprove the existence of such an abnormality. Furthermore, recent studies indicate no defect in red cell membranes when C-HS and non-C-HS animals are compared. Work on lysosomal membranes from C-HS and non-C-HS animals is in progress.[44]

OTHER POTENTIAL USES FOR C-HS ANIMALS

ANTIBIOTIC EVALUATION

Mink, cattle, and man afflicted with the C-HS trait have been shown to be more susceptible to certain bacterial diseases than normal individuals. If the beige mouse should prove to be similarly predisposed to bacterial infections, it may prove useful as a model for initial screening of antibacterial agents. While it is not proposed to eliminate current efficacy-testing procedures of new antibacterial agents, it can be seen that an animal with an increased susceptibility to bacterial organisms would be a sensitive test animal for certain types of antibacterial agents, especially in the early stages of drug testing.

An experiment involving beige and control mice infected with certain organisms, a portion of which have been treated with currently available antibiotics, should indicate if the beige mouse is a more sensitive indicator of the efficacy of the antibiotics than are normal mice. When the exact mechanism of the increased susceptibility of C-HS animals to bacterial diseases is known, it may be possible to develop chemicals that intensify the mechanism of bacterial resistance found in the normal mice by understanding the deficiency in C-HS mice. Man, cattle, and mink would not be suitable experimental animals for this type of procedure, but if the beige mouse, which has the other attributes of the C-HS, proves also to have the increased susceptibility to bacteria, it should be considered as a possible screening animal for testing efficacy of antibacterial preparations.

Adjuvant Evaluation

The efficacy of immunologic adjuvants in achieving higher levels and more durable immunity, when administered with smaller and fewer doses of antigen, has been extensively documented. Despite these findings, adjuvants have found little application in man. The prime reason for this situation has been the lack of studies in animals, particularly long-term chronic toxicity investigations. There is also little information concerning the ultimate fate of the components of the adjuvant administered under the conditions for intended use in man. It seems reasonable to assume that adjuvants combined with tissue antigens administered parenterally at a distant site might be a source of danger to the renal mesangium, with resultant glomerulonephritis. Because of the apparent defective clearance of macromolecular deposits by C-HS mink mesangial cells, it would appear that they may be useful to evaluate adjuvants. Thus, if an abnormal animal, such as one with C-HS, reveals the side effects due to an adjuvant more rapidly and more completely than a normal animal, a major obstacle in the path of the generalized use of adjuvants to enhance the immune response in man might be overcome. Again, the application of such a technique depends upon the demonstration in the beige mouse of a defective glomerular clearance of complexes analogous to that of C-HS mink.

Supported by grants AI 06591 and AI 06477 from the National Institutes of Health, Bethesda, Maryland.

REFERENCES

1. Beguez-Cesar, A. 1943. Neutropenia cronica maligna familiar con granulaciones atipicas de los leucocitos. Boletin de la Sociedad Cubana de Pediatria 15:900–922.
2. Leader, R. W., G. A. Padgett, and J. R. Gorham. 1963. Studies of abnormal leukocyte bodies in the mink. Blood 22:477–484.
3. Padgett, G. A., R. W. Leader, J. R. Gorham, and C. C. O'Mary. 1964. The familial occurrence of the Chediak-Higashi Syndrome in mink and cattle. Genetics 49:505–512.
4. Lutzner, M. A., C. T. Lowrie, and H. W. Jordan. 1967. Giant granules in leukocytes of the beige mouse. J. Hered. 58:299–300.
5. Padgett, G. A. 1960. The Chediak-Higashi Syndrome. Advances in Veterinary Science 12:239–284.
6. Padgett, G. A., C. E. Reiquam, J. B. Henson, and J. R. Gorham. 1968. Comparative studies of susceptibility to infection in the Chediak-Higashi Syndrome. J. Path. Bact. 95:509–522.
7. Lutzner, M. A., J. H. Tierney, and E. P. Benditt. 1965. Giant granules and widespread cytoplasmic inclusions in a genetic syndrome of Aleutian mink: An electron microscopic study. Lab. Invest. 14:2063–2079.
8. Doak, R. A. 1968. Ultrastructural study of the Chediak-Higashi Syndrome in cattle. Masters Thesis. Washington State University.
9. De Robertis, E. O. P. and D. D. Sabatini. 1960. Submicroscopic analysis of the secretory process in the adrenal medulla. Fed. Proc. 19:70–78.
10. Caro, L. G., and G. E. Palade. 1960. Protein synthesis, storage, and discharge in the pancreatic exocrine cell: An autoradiographic study. J. Biol. Chem. 20:473–495.
11. Page, A. R., H. Berendes, J. Warner, and R. A. Good. 1962. The Chediak-Higashi Syndrome. Blood 20:330–343.
12. Padgett, G. A. 1967. Neutrophilic function in animals with the Chediak-Higashi Syndrome. Blood 29:906–915.
13. Windhorst, D. B. 1966. Chediak-Higashi Syndrome: Hereditary gigantism of cytoplasmic organelles. Science 151:81–83.
14. Danes, B. S., and A. G. Bearn. 1967. Cell culture and the Chediak-Higashi Syndrome. Lancet II:65–67.
15. Dent, P. B., L. A. Fish, J. G. White, and R. A. Good. 1966. Chediak-Higashi Syndrome: Observations on the nature of the associated malignancy. Lab. Invest. 15:1634–1642.
16. White, J. G. 1967. The Chediak-Higashi Syndrome: Cytoplasmic sequestration in circulating leukocytes. Blood 29:435–451.
17. White, J. G. 1966. The Chediak-Higashi Syndrome: A possible lysosomal disease. Blood 28:143–156.
18. Padgett, G. A., C. W. Reiquam, R. W. Leader, and J. R. Gorham. 1965. P.A.S. positive material deposited in the R. E. system and neurons of man and animals with the Chediak-Higashi Syndrome. Fed. Proc. 24:493.
19. Kritzler, R. A., J. Y. Terner, J. Lindenbaum, J. Magidson, R. Williams, R. Preisig, and G. B. Phillips. 1964. Chediak-Higashi Syndrome: Cytologic and serum lipid observations in a case and family. Am. J. Med. 36:583–594.

20. Fedorko, M. E., and J. G. Hirsch. 1966. Cytoplasmic granule formation in myelocytes: An electron microscope radioautographic study on the mechanism of formation of cytoplasmic granules in rabbits. J. Cell. Biol. 29:307–316.

21. Wetzel, B. K., R. G. Horn, and S. S. Spicer. 1967. Fine structural studies on the development of heterophil, eosinophil, and basophil granulocytes in rabbits. Lab. Invest. 16:349–382.

22. Bainton, D. F., and M. G. Farquhar. 1968. Differences in enzyme content of azurophil and specific granules of polymorphonuclear leukocytes. I. Histochemical staining of bone marrow smears. J. Cell Biol. 39:286–298.

23. Bainton, D. F., and M. G. Farquhar. 1968. Differences in enzyme content of azurophil and specific granules of polymorphonuclear leukocytes. II. Cytochemistry and electron microscopy of bone marrow cells. J. Cell Biol. 39:299–317.

24. Spicer, S. S., and J. H. Hardin. 1969. Ultrastructure, cytochemistry, and function of neutrophil leukocyte granules. A review. Lab. Invest. 20:488–497.

25. Baggiolini, M., J. G. Hirsch, and C. De Duve. 1969. Resolution of granules from rabbit heterophil leukocytes into distinct populations by zonal sedimentation. J. Cell Biol. 40:529–541.

26. De Duve, C. 1963. The lysosome concept. In Ciba Foundation symposium on lysosomes. Edited by A. V. S. de Reuch and M. P. Cameron. Little, Brown and Co., Boston.

27. Bainton, D. F., and M. G. Farquhar. 1966. Origin of granules in polymorphonuclear leukocytes. Two types derived from opposite faces of the Golgi complex in developing granulocytes. J. Cell Biol. 28:277–301.

28. Spicer, S. S., R. G. Horn, and B. K. Wetzel. 1968. Ultrastructural and cytochemical characteristics of leukocytes in various stages of development. Biochemical Pharmacology Suppl. 143–157.

29. Davis, W. C., W. B. Greene, and S. S. Spicer. 1969. Ultrastructure of bone marrow granulocytes in normal and Aleutian mink. Fed. Proc. 28:36.

30. Lockman, L. A., W. R. Kennedy, and J. G. White. 1967. The Chediak-Higashi Syndrome: Electrophysiological and electron microscopic observations on the peripheral neutopathy. J. Pediat. 70:942–951.

31. Sung, J. H., J. P. Meyers, E. M. Stadlan, D. Cowen, and A. Wolf. 1969. Neuropathological changes in Chediak-Higashi disease. J. Neuropath. and Exp. Neurol. 28:86–118.

32. Donohue, W. L., and H. W. Bain. 1957. Chediak-Higashi Syndrome. A lethal familial disease with anomalous inclusions in the leukocytes and constitutional stigmata: Report of a case with necropsy. Pediatrics 20:416–429.

33. Blume, R. S., J. M. Bennett, R. A. Yankee, and S. M. Wolff. 1968. Defective granulocyte regulation in the Chediak-Higashi Syndrome. New Eng. J. Med. 279:1009–1015.

34. Bennett, J. M., R. S. Blume, and S. M. Wolff. 1969. Characterization and significance of abnormal leukocyte granules in the beige mouse: A possible homologue for Chediak-Higashi Aleutian trait. J. Lab. and Clin. Med. 73:235–243.

35. Windhorst, D. B., J. G. White, A. S. Zelickson, C. C. Clawson, P. B. Dent, B. Pollara, and R. A. Good. 1968. The Chediak-Higashi anomaly and the Aleutian trait in mink: Homologous defects of lysosomal structure. Annals of the New York Academy of Sciences 155:818–846.

36. Windhorst, D. B. 1966. Studies on a hereditary defect involving lysosomal structure. Fed. Proc. 25:358.
37. Root, R. K., R. S. Blume, and S. M. Wolff. 1968. Abnormal leukocyte function in the Chediak-Higashi Syndrome. Clin. Res. 16:335.
38. Zeya, H. I., and J. K. Spitznagel. 1969. Cationic protein-bearing granules of polymorphonuclear leukocytes: Separation from enzyme-rich granules. Science 163:1069–1071.
39. Padgett, G. A., C. W. Reiquam, J. R. Gorham, J. B. Henson, and C. C. O'Mary. 1967. Comparative studies of the Chediak-Higashi Syndrome: Pathology. Am. J. Path. 51:553–571.
40. Henson, J. B., J. R. Gorham, Y. Tanaka, and G. A. Padgett. 1968. The sequential development of ultrastructural lesions in the glomeruli of mink with experimental Aleutian disease. Lab. Invest. 19:153–162.
41. Henson, J. B., J. R. Gorham, G. A. Padgett, and W. C. Davis. 1969. Pathogenesis of the glomerular lesions in Aleutian disease of mink. Arch. Path. 87:21–28.
42. Padgett, G. A. 1969. Aleutian disease virus replication in mink of different genotypes. Fed. Proc. 28:2384.
43. Prieur, D. J. Unpublished data.
44. Holland, J. M. Unpublished data.
45. White, J. G. 1966. Virus-like particles in the peripheral blood cells of two patients with Chediak-Higashi Syndrome. Cancer 19:877–884.

Hereditary Hyperbilirubinemia in Sheep

CHARLES E. CORNELIUS

It is well established that a multiplicity of sequential transport mechanisms exist in the liver for the ultimate biliary excretion of organic anions such as bilirubin, the conjugates of cholic and chenodeoxycholic acids, phylloerythrin, sulfobromophthalein sodium (BSP), and other foreign dyes. Since little is known concerning their transfer from serum albumin through the hepatic plasma membrane, their transport to intracellular conjugation sites and to the biliary canaliculus, and their mechanisms of concentration in the bile, spontaneous mutations occurring in various mammals (man, sheep, and rats) that affect the hepatic transport of bilirubin may aid in the clarification of this complex transport problem. Competition for hepatic uptake, conjugation, or excretion occurring between various organic anions has suggested that certain organic anions may share certain common pathways leading to their biliary excretion.[1]

HEREDITARY NONHEMOLYTIC HYPERBILIRUBINEMIAS IN MAN AND RAT

GUNN RATS

This line of Wistar rats is characterized by an acholuric jaundice transmitted as an autosomal recessive gene.[2] An elevated serum unconjugated bilirubin (UCB) results from a deficiency in hepatic glucuronyl transferase.[3] Electron microscopic examination of liver reveals large areas of agranular endoplasmic reticulum.[4] Bile recovered from Gunn rats is nearly colorless and contains no bilirubin diglucuronide (CB) and only traces of UCB. Despite the inability of the liver to conjugate bilirubin into a soluble polar complex necessary for biliary excretion, these rats maintain a relatively constant serum and tissue UCB concentration by excreting non-diazo-positive polar derivatives of bilirubin into the bile and urine and transferring some UCB across the intestinal mucosa.[5]

CRIGLER-NAJJAR SYNDROME IN MAN

Typical features of this syndrome are high levels of serum UCB (up to 45 mg/100 ml), jaundice by the third day of life, normal conventional liver function test results, no urinary bilirubin, colorless bile with no CB in the bile, and a recessive mode of inheritance. As described previously in the mutant Gunn rat, alternate pathways for bilirubin deposition are present,[5] and no hepatic glucuronyl transferase activity is observed in *in vitro* studies using bilirubin as the receptor.[6] Many infants die before two years of age as a result of kernicterus.[7] Another similar form of unconjugated bilirubinemia in infants[8] has been described, which is characterized by lower concentrations of serum UCB, a possible partial deficiency in hepatic glucuronyl transferase for bilirubin, and a hereditary pattern suggestive of an autosomal dominant gene. Other recent studies by Arias[9] concerning a possible induction of bilirubin conjugation and/or excretion in some infants following phenobarbital administration suggest that a certain etiological heterogeneity exists in hereditary unconjugated hyperbilirubinemia in infants and that further study is warranted.

GILBERT'S SYNDROME IN MAN

Patients with this syndrome, adolescent or adult, exhibit a moderate unconjugated hyperbilirubinemia without other known liver functional or morphological abnormalities.[10] Although defective glucuronide formation has been demonstrated in certain cases both *in vitro* and *in vivo*,[8, 11] there is no consistent agreement, due to the possible heterogeneity of the patients examined. Strong evidence for a diminished hepatic uptake of bilirubin from plasma has been proposed on the basis of plasma clearance studies.[12, 13]

DUBIN-JOHNSON SYNDROME IN MAN

In 1954,[14, 15] a syndrome characterized only by chronic nonhemolytic jaundice and the presence of a brownish-black hepatic pigment was described. In 1948, Rotor and associates[16] described a similar syndrome, but without the hepatic pigment. Elevations in primary CB occur in both the Dubin-Johnson (D-J) and Rotor syndromes. Patients are unaffected but may experience severe jaundice if certain drugs that are known to compete with bilirubin for excretion into the bile are administered. The hepatic excretion (T_m) of BSP in patients with this syndrome is greatly depressed while the relative hepatic storage of BSP is normal.[17] The inheritance of the Dubin-Johnson syndrome is not yet clear but represents a defect in the transfer of CB and certain other organic anions from liver cells into bile.

HEREDITARY NONHEMOLYTIC HYPERBILIRUBINEMIAS IN SHEEP

Hepatic diseases are abundant in animals and truly represent "experiments in nature." Many unique liver diseases of animals either have been only superficially studied or still remain unrecognized. Investigations continue to produce hepatic lesions in the commonly used laboratory animals and, in most instances, unsuccessfully attempt to mimic counterpart human hepatopathies. It would seem more advisable and rewarding for clinical scientists to utilize animal models with spontaneously occurring liver diseases. Such animals are recognized by the veterinary medical community and are often available for study.[18]

Sheep are a unique species in which to detect hepatic transport defects. Cutaneous photosensitivity occurs in ruminants because of the inability of a defective liver to excrete phylloerythrin, a porphyrin that results from the ingested green feed. Phylloerythrin is formed from chlorophyll by enteric bacteria and is quantitatively extracted from the portal blood and excreted into the bile under normal conditions by the liver. Any hepatic disease in sheep, whether acquired as a result of the ingestion of hepatotoxins or inherited as an organic anion transport defect for bilirubin, phylloerythrin, etc., may result in defective hepatic uptake and/or excretion of such organic anions and in subsequent clinical photodynamic dermatitis and ultimate death. Lambs exhibiting photosensitivity at the time of weaning and their first ingestion of green grass are prime candidates for such inherited hepatic transport disorders.

DUBIN-JOHNSON (D-J) SYNDROME IN CORRIEDALE SHEEP: AN ORGANIC ANION EXCRETORY DEFECT

CLINICAL OBSERVATIONS

Acute photosensitization was first observed in these mutants on a ranch near Sacramento, California, in 1963.[19] Photophobia and ulcerations on the ears and around the eyes were characteristically seen in weanling male and female lambs after the first ingestion of grass. A progressive loss of weight from anorexia occurred prior to death and established this syndrome to be a lethal trait in sheep under field conditions. No clinical jaundice was observed. When affected lambs were promptly removed from sunlight, they commenced to recover. The mode of inheritance still remains unclear, although a D-J offspring from a nonaffected ewe has been recorded.

PATHOLOGIC OBSERVATIONS

Livers were observed at necropsy to be grossly normal in all respects except for a brownish-black discoloration. Microscopic examination revealed dark pigment granules that were pericanalicularly distributed in the hepatic cells. Ultrastructural studies subsequently showed that the pigment was localized in hepatic lysosomes, as previously reported by Essner and Novikoff[20] in human cases of D-J syndrome. The renal cortices of two D-J ovine mutants were also discolored with brownish-black pigment and clearly discernible in the proximal and distal convoluted tubules. A recent study has demonstrated that no hepatic pigment is present at birth, but that it slowly accumulates in the liver until amounts characteristic of adults present at 5 to 6 months of age.[21]

Human and ovine hepatic pigments were isolated and studied by Arias et al.[22] using sucrose density gradient centrifugation and various physicochemical procedures. Isolated pigments from man and sheep gave identical results with elemental analysis, solubility in organic solvents, ultraviolet absorption spectra, and electron spin resonance characteristics.[22, 23] These detailed studies suggested that the pigments are melanins and may possibly arise from defective excretion of metabolites of tyrosine, tryptophan, or phenylalanine. Bile and urine were collected from normal and mutant sheep after the injection of epinephrine-7-^{3}H. Mutant sheep excreted 1 percent of the administered radioactivity into the bile as compared to 18 percent in normal sheep. Incorporation of radioactivity occurred by the third day and was followed by a very slow turnover rate.[23] The hepatic pigment is postulated to result, at least in part, from impaired hepatic excretion of epinephrine metabolites that are oxidized to insoluble polymers of melanin and stored subsequently in the pericanalicular dense bodies.

Such other similar pigmentations of the liver as pigmentary liver disease in howler monkeys in Argentina[24] and the "environmental lipofuscinosis" in sheep ingesting mulga leaves (*Acacia aneura*) in Australia[25] are not known to be associated with any anion excretory defect as in the D-J syndrome.

PHYSIOLOGIC OBSERVATIONS

Dubin-Johnson syndrome in man has been established as a chronic familial nonhemolytic jaundice in which conjugated bilirubin (CB) accumulates in the plasma due to an excretory defect for CB in the liver cell. Recent studies on both human and Corriedale ovine mutants indicate that a decreased hepatic excretory maximum exists for bilirubin and BSP. Iopanoic acid, indocyanine green, and phylloerythrin are also not excreted maximally.[19] Corriedale mutants, when infused with bilirubin, excrete bilirubin maximally at only 5 percent of the rate of normal sheep. Maximal concentrations of bilirubin in the bile of infused mutants were 1.5 mg/ml as compared to 16 mg/ml on average in normal infused sheep.[26]

The plasma of mutants contains significantly higher concentrations of total bilirubin (1.0 ± 0.2 mg%) than does that of normal lambs (0.4 ± 0.02 mg%) at birth. CB in all cases constituted over 60 percent of the total serum bilirubin levels. Differences are even greater at six months of age (mutants, 1.6 ± 0.1, and normals, 0.2 ± .02). Plasma clearance of BSP (5 mg/kg) in mutant neonates was, on average, 5 ± 1 min ($T_{1/2}$) as compared to 3 ± 0.4 min in normal lambs. Half-time values for BSP disappearance in mutants increased progressively and at six months of age were on average 33 ± 1 min, in contrast to the values in normal sheep, which remained the same.[21] Relative hepatic storage (S) of BSP was normal in mutants, whereas the hepatic T_m for BSP was markedly decreased[22] when measured by the indirect technique as described by Wheeler *et al.*[27] The BSP excretory rate was measured in the bile of one mutant following its intravenous injection and never exceeded 7 percent of that observed in normal sheep.[22] The concentrations of phylloerythrin in serum and liver of mutants were over 15 times greater than in normal sheep, whereas phylloerythrin concentrations in mutant bile were one-seventh of that observed in normal sheep.[22]

Previous studies on the hepatic excretion of various organic anions have suggested that all such anions might share a common hepatic excretory pathway. A unique opportunity existed to determine if bile acids (taurocholate) shared a common excretory mechanism with BSP. Following the infusion of taurocholate, it was observed to be excreted normally in both mutant and control sheep.[28] Taurocholate infusion enhanced the maximum hepatic excretion of BSP in normal sheep, but not in mutants. These studies suggest dissimilarities in hepatic excretory steps for certain organic anions. It was also of interest that the excretion of procaine amide ethobromide, an organic cation, was not different in mutant and normal sheep.[28]

Since all studies on mutant Corriedale sheep appear morphologically and functionally similar to those in human cases of D-J syndrome, it would appear this ovine mutant should be useful in studying both the human disease and hepatic excretory function of organic anions.

CONGENITAL PHOTOSENSITIVITY AND HYPERBILIRUBINEMIA IN SOUTHDOWN SHEEP: AN ORGANIC ANION UPTAKE DEFECT

Recent unpublished data from our laboratory[29] suggest this inherited bilirubin defect resembles Gilbert's syndrome in man. This disorder was first observed in New Zealand sheep in 1942[30] and proved to be inherited as a single autosomal recessive factor.[31] Photosensitivity results from retention of phylloerythrin. Mutants do not exhibit jaundice, but have a hyperbilirubinemia of between 0.5 and 2 mg%, of which over 60 percent is UCB. Preliminary studies in our laboratories[29] used bilirubin–[14]C in tracer quantities in mutant Southdown sheep to determine the plasma bilirubin–[14]C disappearance rates. From these studies, a multicompartmental model for UCB similar to that of Barrett *et al.*[13] was used for computer analysis. Kinetic studies revealed more UCB in a "rapidly mixing pool" and a similar amount in the "storage pool." The rapidly mixing pool, which is primarily the plasma pool, contained 2½ times more UCB in mutants than in normal Southdown sheep. In addition, a smaller fraction of the rapidly mixing pool was transported per minute into the storage pool, which has been postulated to be primarily the liver. The total excretory rate of UCB from both combined pools is similar and is proportional at a different steady state to the total UCB pool.[29] The bilirubin transport defect in Southdown mutants was similar to that observed by Barrett et al.[13] and Billing et al.[32] in Gilbert's syndrome in man.

It was of special interest in the Southdown mutant that unlike hepatic function in Gilbert's syndrome in man, the hepatic uptake of BSP, rose bengal, indocyanine green, and cholate is also defective. After the intravenous injection of BSP into mutant Southdown sheep, a single exponential component occurred ($T_{1/2}$ on average of 43 min as compared with 3 min in normal sheep) representing the disappearance of dye without the rapid uptake or mixing component observed in clearance curves in normal sheep.[33] No BSP-glutathione conjugates were observed in the plasma BSP pool 45 min after the injection of unconjugated BSP. BSP recovered from the bile of a mutant was 70–80 percent conjugated as in normal sheep. Relative hepatic storage (S) and transport maxima (T_m) for BSP in two mutants as calculated by the indirect method of Wheeler et al.[27] were 0 mg/mg/100 ml/kg and 0.025 mg/min/kg, respectively. S and T_m in normal Southdown sheep are 1.8 ± 0.9 mg/mg/100 ml/kg and 0.24 ± 0.02 mg/min/kg, respectively. By the direct collection of fistula bile from a mutant, it was observed that 4 percent of the normal excretory maxima of BSP was excreted per minute.

Serum albumin binding of BSP was determined to be normal in mutants as measured using sephadex gel filtration and electrophoretic analysis. Cross-transfusion studies allowing for a complete admixture of a normal and mutant sheep prior to BSP clearance studies also did not suggest a defect in the binding of BSP by mutant serum albumin.

PATHOLOGIC FINDINGS

Although this lethal trait in mutants under field conditions can be suppressed by preventing photosensitivity, three mutants have died recently from an unexplained renal fibrosis and subsequent uremia. Preliminary experiments suggest that mutants lack the ability to concentrate urine.[34]

Results of preliminary histologic studies of necropsy specimens by both light and electron microscopy revealed normal hepatic structure. No increase in pigment-containing lysosomes as in the D-J syndrome was observed.

SUMMARY

Because there is little documentation of possible biochemical mechanisms and energy dependence in organic anion transport, ovine mutants with hyperbilirubinemia should be useful biomedical models in which to study the active transport of bilirubin by the liver. Similarities of Corriedale mutants to the Dubin-Johnson syndrome and of Southdown mutants to Gilbert's syndrome should allow for new experimental approaches to study the defective mechanisms involved in these two human hepatopathies.

These studies were supported in part by USPHS Grant No. AM-11384-02.

REFERENCES

1. Schenker, S., and R. Schmid. 1964. Excretion of C^{14}-bilirubin in newborn guinea pigs. Proc. Soc. Exp. Biol. Med. 115:446–448.
2. Gunn, C. K. 1938. Hereditary acholuric jaundice. J. Hered. 29:137–139.
3. Arias, I. M. 1959. A defect in microsomal function in nonhemolytic acholuric jaundice. J. Histochem. Cytochem. 7:250–253.
4. Novikoff, A. B., and E. Essner. 1960. The liver cell. Amer. J. Med. 29:102–131.
5. Schmid, R., and L. Hammaker. 1963. Metabolism and disposition of C^{14}-bilirubin in congenital nonhemolytic jaundice. J. Clin. Invest. 42:1720–1734.
6. Szabo, L., Z. Kovacs, and P. B. Ebrey. 1962. Crigler-Najjar syndrome. Acta Pediat. Hung. 3:49–70.
7. Crigler, J. F., and V. A. Najjar. 1952. Congenital familial nonhemolytic jaundice with kernicterus. Pediatrics 10:169–180.
8. Arias, I. M. 1962. Chronic unconjugated hyperbilirubinemia without overt signs of hemolysis in adolescents and adults. J. Clin. Invest. 41:2233–2245.
9. Arias, I. M. 1969. Personal communication.
10. Lester, R., and R. F. Troxler. 1969. Recent advances in bile pigment metabolism. Gastroenterology 56:143–169.
11. Arias, I. M., S. Wolfson, J. F. Lucey, and R. J. McKay. 1965. Transient familial neonatal hyperbilirubinemia. J. Clin. Invest. 44:1442–1450.
12. Howatson, A. F., and A. W. Ham. 1955. Electron microscope study of sections of two rat liver tumors. Cancer Res. 15:62–69.
13. Barrett, P. V. D., P. D. Berk, M. Menken, and N. I. Berlin. 1968. Bilirubin turnover studies in normal and pathologic states using bilirubin-[14]C. Ann. Intern. Med. 68:355–377.
14. Dubin, I. N. 1958. Chronic idiopathic jaundice: Review of fifty cases. Amer. J. Med. 24:268–292.
15. Sprinz, H., and R. S. Nelson. 1954. Persistent nonhemolytic hyperbilirubinemia associated with lipochrome-like pigment in liver cells: Report of four cases. Ann. Intern. Med. 41:952–962.

16. Rotor, A. B., L. Manahan, and A. Florentin. 1948. Familial nonhemolytic jaundice with direct van den Bergh reaction. Acta Med. Phillipina 5:37–49.
17. Arias, I. M. 1961. Studies of chronic familial nonhemolytic jaundice with conjugated bilirubin in the serum with and without an unidentified pigment in the liver cells. Amer. J. Med. 31:510–518.
18. Cornelius, C. E. In press. Hepatic diseases in animals. *In* H. Popper and F. Schaffner [ed] Progress in liver diseases, Vol. 3.
19. Cornelius, C. E., I. M. Arias, and B. I. Osburn. 1965. Hepatic pigmentation with photosensitivity: A syndrome in Corriedale sheep resembling Dubin-Johnson syndrome in man. J. Amer. Vet. Med. Ass. 146:709–713.
20. Essner, E., and A. B. Novikoff. 1960. Human hepatocellular pigments and lysosomes. J. Ultrastructure Res. 3:374–391.
21. Cornelius, C. E., R. R. Gronwall, B. I. Osburn, and G. H. Cardinet. 1968. Dubin-Johnson syndrome in immature sheep. Amer. J. Dig. Dis. 13:1072–1076.
22. Arias, I. M., L. Bernstein, R. Toffler, C. E. Cornelius, A. B. Novikoff, and E. Essner. 1964. Black liver disease in Corriedale sheep: A new mutation affecting hepatic excretory function. J. Clin. Invest. 43:1.
23. Arias, I. M., L. Bernstein, R. Toffler, and J. Ben Ezzer. 1965. Black liver disease in Corriedale sheep: Metabolism of tritiated epinephrine and incorporation of isotope into the hepatic pigment *in vivo*. J. Clin. Invest. 44:1026.
24. Katz, S., A. Gilardoni, N. Genovese, R. W. Wikinski, C. E. Cornelius, and M. R. Malinow. 1968. Liver function studies in free ranging howler monkeys with hepatic pigmentation. Lab. Animal Care 18:626–630.
25. Winter, H. 1961. An environmental lipofuscin pigmentation of livers. Univ. of Queensland Papers, Santa Lucia, Australia, 1:5–71.
26. Upson, D. W., and C. E. Cornelius. Unpublished data. Dept. of Physiological Sciences, Kansas State Univ., Manhattan.
27. Wheeler, H. O., J. I. Meltzer, and S. E. Bradley. 1960. Biliary transport and hepatic storage of sulfobromophthalein sodium in the unanesthetized dog, in normal man and in patients with hepatic disease. J. Clin. Invest. 39:1131–1144.
28. Alpert, S., M. Mosher, A. Shanske, and I. M. Arias. 1969. Multiplicity of hepatic excretory mechanisms for organic anions. J. Gen. Phys. 53:238–247.
29. Mia, A. S., R. R. Gronwall, and C. E. Cornelius. Unpublished data. Dept. of Physiological Sciences, Kansas State Univ., Manhattan.
30. Cunningham, I. J., C. S. M. Hopkirk, and J. F. Filmor. 1942. Photosensitivity diseases in New Zealand. I. Facial eczema: Its clinical, pathological and biochemical characteristics. N.Z. J. Sci. Tech. 24A:185–198.
31. Hancock, J. 1950. Congenital photosensitivity in Southdown sheep. A new sublethal factor in sheep. N.Z. J. Sci. Tech. 32A:16.
32. Billing, B. H., R. Williams, and T. G. Richards. 1964. Defects in hepatic transport of bilirubin in congenital hyperbilirubinemia: An analysis of plasma bilirubin disappearance curves. Clin. Sci. 27:245–257.
33. Cornelius, C. E., L. W. Holm, and D. E. Jasper. 1958. Bromsulphalein clearance in normal sheep and in pregnancy toxemia. Cornell Vet. 48:306–312.
34. McGavin, M. D., R. R. Gronwall, A. S. Mia, G. W. Osbaldiston, and C. E. Cornelius. Unpublished data. Depts. of Infectious Diseases, Pathology and Physiological Sciences, Kansas State Univ., Manhattan.

Animal Models of Atherosclerosis

THOMAS B. CLARKSON, ROBERT W. PRICHARD,
BILL C. BULLOCK, NOEL D. M. LEHNER,
HUGH B. LOFLAND, and RICHARD W. ST. CLAIR

Atherosclerosis has been defined by the World Health Organization as
a "variable composition of intimal changes in which there is focal accumulation of lipids, complex carbohydrates, blood and its products,
fibrous tissue and calcium deposits, and associated with medial
changes." Among North American human beings atherosclerosis and
its sequele are the leading causes of morbidity and mortality. For this
reason, a very large effort has been made to obtain a better understanding of the pathogenesis of this disease and, through a better understanding of pathogenesis, to be in a position to approach prophylaxis
and therapy on a rational basis. Since it is very difficult to determine
the precise extent of atherosclerosis in patients, the principal research
effort has been directed towards experiments on animal models of the
disease. The purpose of this communication is to summarize our
thoughts about some of the animal models that are available currently
for the study of atherosclerosis. We shall try not to wander from this
subject although it is tempting to include something about animal
models of myocardial infarction in the hopes of stimulating research
in that area. Very little has been done about producing myocardial
infarction experimentally by the means through which it usually occurs in man—that is, complications of coronary artery atherosclerosis.

The suitability of animal models for research on atherosclerosis has
been, and continues to be, one of the most controversial subjects in
comparative and experimental pathology. The controversy may arise
from the desire by some workers to find an animal model that would
be ideally suited for all types of research on the disease. It seems evident that no single species can play this role. It seems more reasonable

to expect that animal models will evolve as most suitable for studies on such particular facets as cholesterol absorption, cholesterol excretion, aortic atherosclerosis, coronary artery atherosclerosis, cerebral artery atherosclerosis, and peripheral arterial disease. A great deal of work remains to be done to fully characterize the animal models now in use. Until we have a better understanding of existing animal models, sweeping recommendations concerning human beings should be made cautiously.

Perhaps it would be well to approach the subject historically, beginning with the early Russian animal experiments. Consideration of animal models of atherosclerosis largely dates from Anitschkow's work[3] begun over 50 years ago. Stimulated by earlier attempts to provoke atherosclerosis through feeding fats and egg yolk, he gave rabbits cholesterol-containing diets, which produced accumulations of lipids in their aortic intima. These yellow lesions looked like those of human aortic atherosclerosis, although of slightly different distribution, occurring more in the proximal aorta than in the distal. The epicardial arteries of rabbits are not much affected by lipid accumulations, but the small intramyocardial branches do become diseased. Complications of atherosclerosis such as hemorrhage into the plaque, ulceration and thrombus formation do not occur without additional extensive manipulations such as administration of epinephrine and vitamin D, and alternating feedings of cholesterol-containing and normal diet.[11] Most strikingly different from the human situation, however, are factors that are often overlooked in assessing the cholesterol-fed rabbit—namely the striking hypercholesterolemia of roughly 40 times the normal rabbit level, and the appearance of lipid in almost all of the organs and tissues of the body.[35] After five months or so of cholesterol feeding the rabbits begin to lose weight, and, when autopsied, have lipid deposits in such places as the kidney, ciliary body, the adrenal cortex, the ovary, the bowel mucosa, the lymphatic nodules of the viscera, the footpads, and many other places. The appearance is that of a storage disease, and the total situation presents only a very superficial resemblance to human atherosclerosis.

The long-term fate of diet-induced atherosclerotic lesions in rabbits has not been studied in detail previously, although Prior and Ziegler[34] have reported on some short-term regression studies. In our laboratories we fed New Zealand White rabbits a diet containing 10 percent fat and 0.5 percent cholesterol for one year. At the end of that year some of the animals were autopsied to determine the extent and type of atherosclerotic lesions that had developed. The diet-induced lesions were typical of the ones that have been described. The diet of the remaining animals was changed to rabbit "chow" (containing 3 percent fat and no cholesterol); they were fed this diet for the next three years and then autopsied. Atherosclerosis was still present in all of the animals although no visceral lipidosis remained. The morphologic characteristics of the lesions bore a much closer resemblance to human atherosclerosis, both grossly (Figure 1) and microscopically (Figures 2–5), than those examined at the end of one year. It may be that rabbits that are first fed cholesterol, and then returned to normal diets until visceral lipidosis has regressed, are more suitable as animal models.

Perhaps the next animal to receive major attention in atherosclerosis research was the chicken, largely through the efforts of Katz, Stamler, and Pick in Chicago.[17] Chickens develop aortic atherosclerosis spontaneously,[12, 30] but little use has been made of the nonmanipulated chicken. The cholesterol-fed cockerel develops both aortic and coronary lipid deposits quite readily, and has been used especially to show that the administration of estrogens ameliorates coronary artery disease, while having little or no effect on the aortic lesions.[31] On the basis of this work, coupled with clinical observations on the alleged effects of castration in eliminating the protection from atherosclerosis that ovaries are said to give women, trials of estrogen in human patients have been carried out; the results are conflicting. Currently a search is

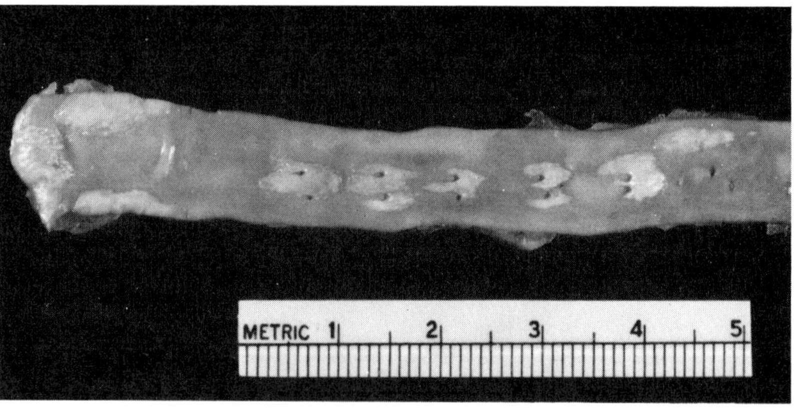

FIGURE 1 The thoracic aorta from a rabbit fed a cholesterol-containing diet for one year and stock diet without cholesterol for three years has plaques in the arch and around the origins of the intercostal arteries. ×1.2

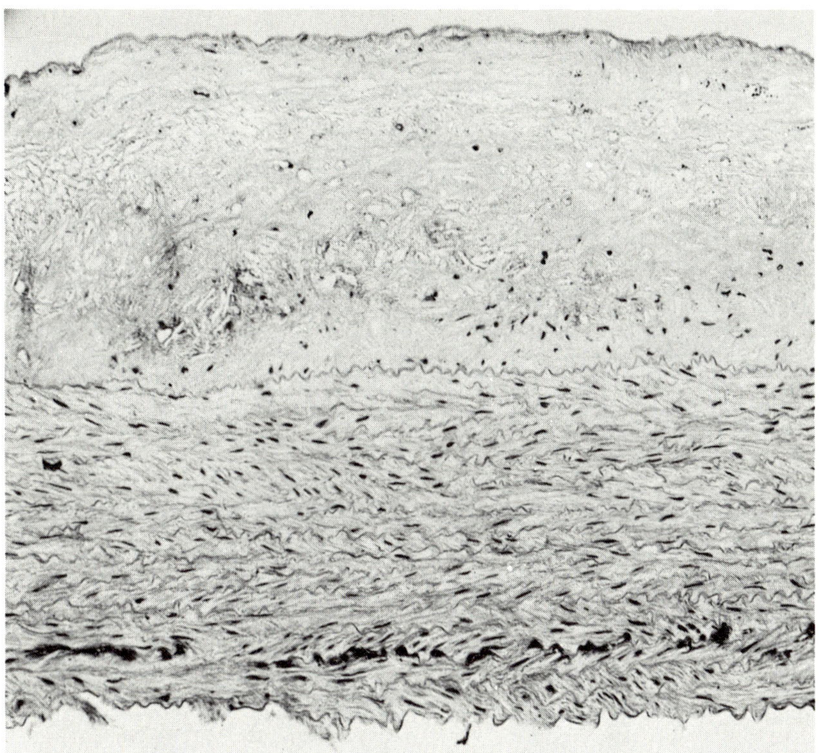

FIGURE 2   A photomicrograph of the aorta shown in Figure 1. The intima is about the same thickness as the media and there is some calcification of tissue adjacent to lipid clefts. The internal elastic membrane is relatively intact. H & E ×113

on for steroids that will have a favorable effect on lipid deposits, without affecting secondary sexual characteristics.

Rats appeared on the research scene as heralds of the view that the best animal for atherosclerosis research is one that not only does not get the disease while on its usual diet, but that is resistant to induction of the disease. Supporters of this view feel that such an animal will most clearly show the effect of a given manipulation, unclouded by the action of some naturally-occurring, undefined agency. The chief focus of rat experiments has been the phase of human atherosclerosis characterized by necrosis within the lesions, and thrombus formation overlying them. To produce fat-containing arterial lesions rats are

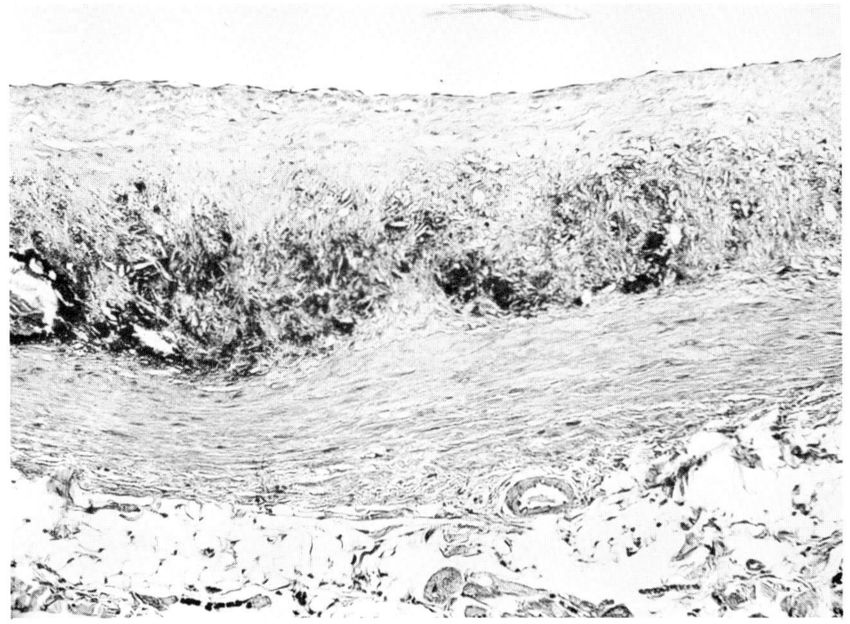

FIGURE 3   A photomicrograph of the aorta just above the coronary ostia of the same rabbit. Calcium deposits are present at the base of the thickened intima near an indistinct internal elastic membrane. H & E ×106

given diets containing 40 percent fat, 5 percent cholesterol (human diets contain about 0.02 percent), propylthiouracil, and a variety of such other substances as choline and sodium cholate.[44] In some experiments investigators have depressed the thyroid function by ablation rather than by feeding the anti-thyroid agent. Lipid-containing lesions have developed in the aorta, and thrombosis has occurred in other vessels, but not in relation to atherosclerotic lesions. The rat experiments have shown that butter is especially provocative of thrombosis as compared with other fats.

The development of special strains of rats with hypo- and hypercholesterolemia has been reported recently.[1] The availability of these strains is a significant advance in the use of animal models for investigating mechanisms of genetic control of cholesterol homeostasis.

During the past decade, nonhuman primates have become of increasing interest as animal models of atherosclerosis. This interest has developed because it is believed that data obtained from these animals may be more directly applicable to atherosclerosis in human beings than that obtained from lower animals. It now appears that most species of nonhuman primates have some degree of naturally occurring atherosclerosis, although clinical complications of the disease among nonhuman primates in their natural environment have not been observed. Among the primates, the natural occurrence of atherosclerosis among New World monkeys has been studied more extensively than that among the Old World species. During the summers of 1964 and 1965, workers from our laboratories, in collaboration with members of the Department of Pathology of the Louisiana State University School of Medicine, surveyed the naturally occurring arterial lesions of six species of free-living New World monkeys in the vicinity of Leticia, Colombia (Table 1). The prevalence and extent of aortic atherosclerosis found in these monkeys have been published by Lofland et al.[20] Among the species studied, squirrel monkeys (*Saimiri sciureus*) had both the highest prevalence and the greatest extent of aortic atherosclerosis. Considerably less atherosclerosis was found among the *Lagothrix* sp., *Cebus* sp., and *Saguinus nigricollis*, although the *Lagothrix* are primarily juvenile animals, and it is not justifiable to compare them with the adults of the other species.

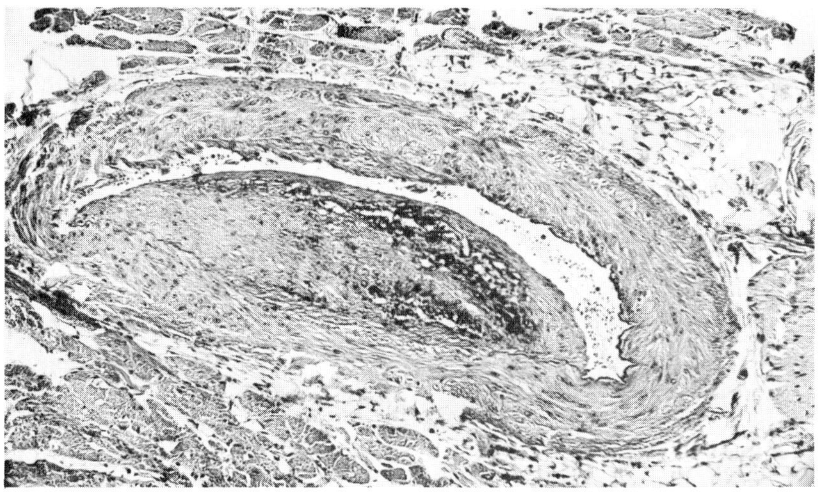

FIGURE 4   A plaque in a coronary artery of the same rabbit is present in this photomicrograph. The plaque is mostly connective tissue and calcium salts and apparently contained only a small amount of lipid. H & E ×91

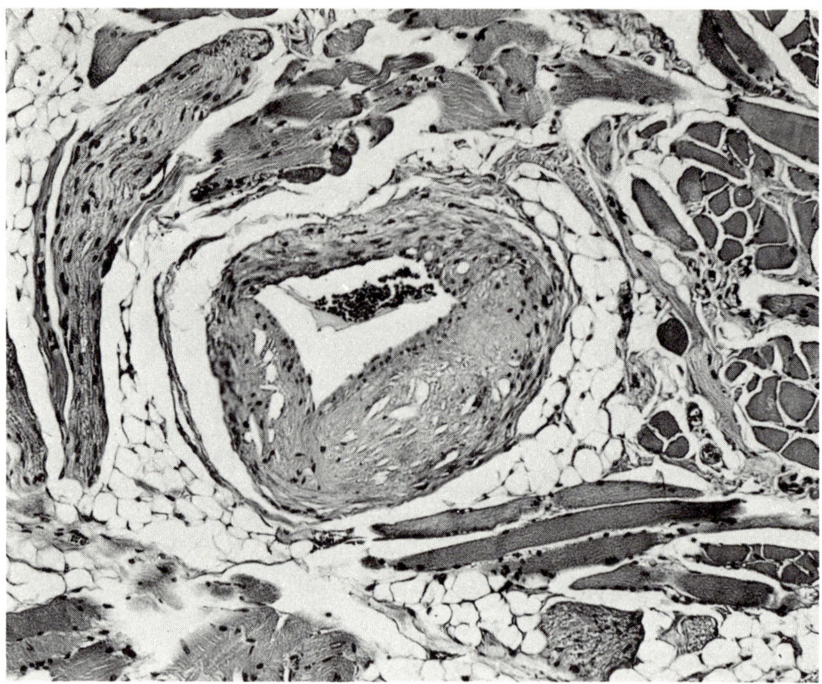

FIGURE 5   A photomicrograph of an artery in the tongue of the same rabbit. The intima is thickened and sclerotic. The media is thinned where the intima is thickest. H & E ×115

TABLE 1   Total Serum Cholesterol Levels and Naturally Occurring Aortic Atherosclerosis of New World Monkeys[a]

Species	No.	Total Serum Cholesterol[b] (mg/100 ml)	Atherosclerosis Index[c]	
			Thoracic	Abdominal
Ateles spp.	29	130 ± 7.0	1.88 ± 0.6	0.60 ± 0.3
Saimiri sciureus	220	105 ± 2.1	4.95 ± 0.7	2.69 ± 0.5
Cebus albifrons	57	90 ± 3.6	1.88 ± 0.6	0.96 ± 0.5
Cebus apella	21	98 ± 10.7	0.0	0.0
Saguinus nigricollis	44	69 ± 3.5	0.65 ± 0.5	0.0
Lagothrix spp.	44	133 ± 12.1	0.38 ± 0.3	0.63 ± 0.5

[a] Adapted from data of Lofland *et al.* 1967. Arch. Path. 83:211.
[b] Expressed as mean for the group followed by standard error of the mean.
[c] Expressed as mean percent of intimal surface that was sudanophilic followed by standard error of the mean.

Naturally occurring atherosclerosis among free-living howler monkeys (*Alouatta caraya*), has been studied by Malinow and Maruffo.[22] The extent of lesions in the aorta of these animals increased with age from 0.25 percent of the aortic surface being involved among infants to 7.5 percent among adult males.

Among the Old World nonhuman primates, relatively few species have been studied in sufficient depth for one to be certain about the prevalence and extent of naturally occurring atherosclerosis. On the basis of excellent pathologic studies by Finlayson,[15] Lindsay and Chaikoff,[19] and many others, on the small number of apes and Old World monkeys dying in zoos, it is apparent that naturally occurring disease is not infrequent in this group. Baboons (*Papio doguera*) have been carefully studied to determine the prevalence and extent of their naturally occurring arterial lesions. The studies by McGill and his co-workers[27] showed that baboons have a high prevalence of aortic fatty streaking, but little or no coronary artery atherosclerosis. In a study of wild-living gray-cheeked mangabeys, Strong[39] also found naturally occurring aortic fatty streaks and, in this species, some coronary artery lesions. A recent study on arterial lesions among wild-caught rhesus monkeys has been reported by Chawla and his coworkers.[6] In this study, 150 rhesus monkeys were examined. It would appear that naturally occurring atherosclerosis is infrequent in this species since fatty streaks of the aorta were seen in only four of the animals. The rarity of naturally occurring atherosclerosis seen in this study is in agreement with the observations of the group at the Russian Primate Center at Sukhumi.

Diet-induced atherosclerosis has been studied in certain of the apes and Old World monkeys, particularly chimpanzees (*Pan troglodytes*),[2] baboons (*Papio* spp.),[16, 40] rhesus monkeys (*Macaca mulatta*),[37, 38, 41–43, 46, 47] and cynomolgus macaques (*Macaca irus*).[18] Among the New World monkey species, diet-induced atherosclerosis has been studied in squirrel monkeys (*Saimiri sciureus*),[23, 28, 32] woolly monkeys (*Lagothrix lagothricha*),[32] and cebus monkeys (*Cebus albifrons*).[4, 5, 8, 9, 25, 32, 45]

The lesions observed in the chimpanzees bore a striking resemblance to human atherosclerosis, particularly in the cerebral arteries. However, the high cost of these animals, along with their scarcity, makes their general use in atherosclerosis both impractical and unwise.

Considerable attention has been given to the possible usefulness of baboons as animal models of atherosclerosis. It seems unlikely that they will occupy a prominent position among the primate models for studies on lesions. In response to cholesterol-containing diets, they do develop fatty streaks in the aortic intima, although these lesions are usually very small. Coronary artery atherosclerosis is almost never seen and the few coronary lesions noted have been inconsequential. It does appear likely, however, that baboons may be of particular value as animal models for the study of whole-body cholesterol metabolism. It appears that they share with man a limited ability for intestinal absorption of cholesterol, and that the limited amount absorbed is excreted primarily as neutral steroids.

Rhesus monkeys are commonly used as animal models of atherosclerosis. Their principal advantage is the rather severe extent to which these animals develop atherosclerosis when fed cholesterol-containing diets. The severity of the lesions in this species exceeds that in all other species we have studied. These animals not only develop extensive aortic atherosclerosis, but extensive coronary artery and cerebral artery atherosclerosis as well. The anatomic distribution of the coronary artery lesions in these animals is another of their advantages, the lesions being most significant in the proximal branches of the coronary arteries that are epicardially distributed (Figure 6). In our experience, renal artery atherosclerosis has been a significant lesion in this species (Figures 7 and 8), and the possibility that atherosclerotic rhesus monkeys would develop renal hypertension seems real.

Squirrel monkeys comprise the primate species with which we have had the most experience in our laboratories. These animals offer a number of advantages as animal models of atherosclerosis; they are relatively inexpensive, are easily caged and manipulated, breed well in

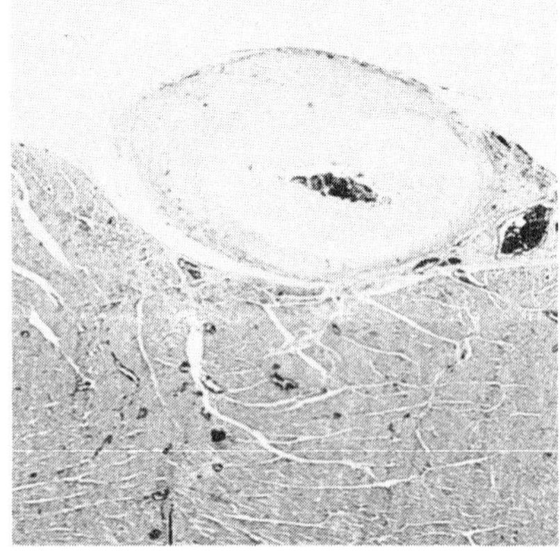

FIGURE 6   The epicardial coronary artery of a rhesus monkey has a greatly thickened intima which reduced the apparent lumen to about 20 percent of its original size. H & E X18

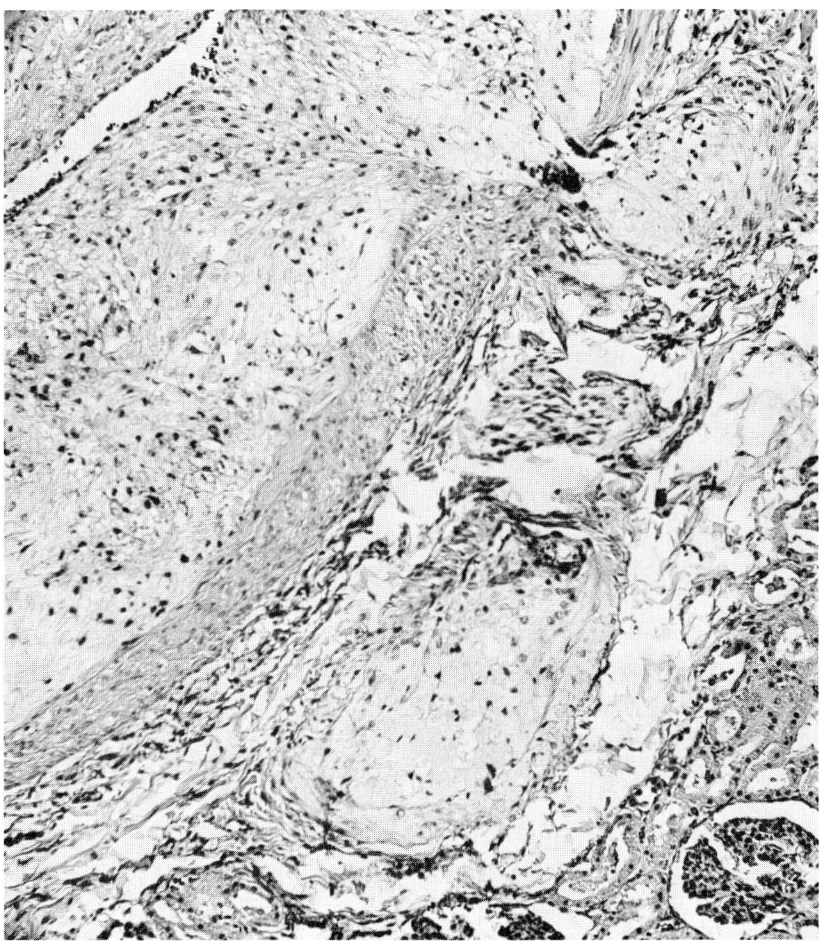

FIGURE 7   The original lumen of the renal artery of a rhesus monkey has been reduced to the slit seen at the upper left of the photomicrograph. There is calcification of the origin of a small branch (upper right) and the lumen of another small branch (between the larger artery and renal tissue) is almost completely filled with "foam cells." H & E ×118

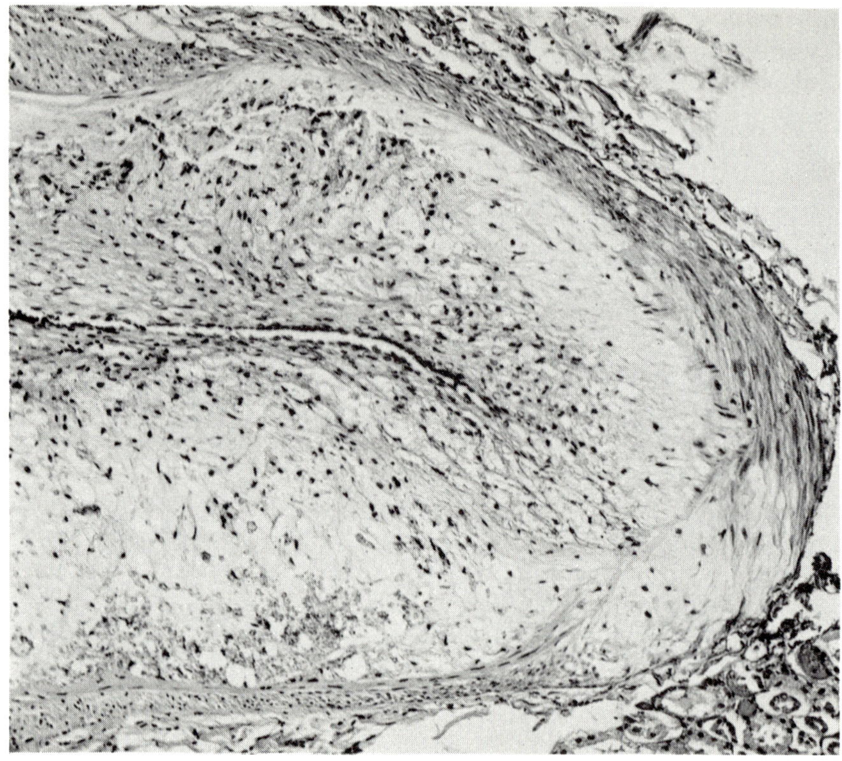

FIGURE 8   The renal artery of a rhesus monkey has large atheromatous deposits with some mineralization central to the internal elastic membrane (bottom of photomicrograph). The media at the lower right has a clear appearance due to the removal of lipid during preparation of the paraffin section. H & E ×113

captivity, develop fairly extensive atherosclerosis, and share with man several important aspects of whole body cholesterol metabolism.

It appears that squirrel monkeys may be particularly valuable for studies on the effect of estrogens on cholesterol metabolism. Over a 25-month period we have studied the serum cholesterol levels of squirrel monkeys fed a semisynthetic diet containing 25 percent by weight of one of three types of dietary fat and 1 mg per calorie of cholesterol. A striking reduction of serum cholesterol levels occurred among the female monkeys during the breeding season (March through June). Table 2 summarizes the serum cholesterol levels of these monkeys before and during the breeding season. The changes in serum cholesterol levels appeared to be the same whether or not the animals became pregnant during the breeding season. We assume that these differences in cholesterol metabolism are the result of increased estrogenic activity; as yet, however, we have no experimental data to support this assumption.

Perhaps the most important characteristic of squirrel monkeys, used as animal models for research on atherosclerosis, is their striking individuality in response to cholesterol-containing diets. Among squirrel monkeys consuming the same atherogenic diet (containing 1 mg of cholesterol per calorie of diet), the serum cholesterol level of some monkeys increases only slightly above baseline levels (200–300 mg/ml) while for others it increases to relatively high hypercholesterolemic levels (800–1,000 mg/100 ml). We have designated these monkeys as hypo- and hyper-responders and, by selective breeding, have developed strains of squirrel monkeys with these metabolic characteristics. We feel that the availability of these strains of monkeys is important for future research on mechanisms of genetic control of cholesterol homeostasis.

Diet-induced atherosclerosis of cebus monkeys is characterized by slight lesions in the aorta; relatively extensive lesions in the coronary arteries of adult but not of young monkeys; rather large, raised plaques at the carotid artery bifurcation; lesions of the small intramuscular arteries of the tongue, the uterus, and kidneys; and the absence of lesions in the cerebral vessels. *Cebus albifrons* monkeys have been of particular interest as animal models because of a rather striking age difference in the susceptibility of these animals to diet-induced atherosclerosis. We found that young, immature cebus monkeys, consuming the atherogenic diet (25 percent lard, 0.5 percent cholesterol), maintained lower levels of serum cholesterol than did adults fed the same diet, with adult male monkeys having markedly higher levels than adult females. No difference was detectable among control animals of either sex and age group. The differences in serum cholesterol between the young and adult cholesterol-fed monkeys could not be accounted for by differences in consumption, absorption, or hepatic synthesis of cholesterol in studies conducted after one year of feeding. The young monkeys seemed to compensate for dietary cholesterol by enhanced excretion, as suggested by a somewhat faster disappearance of intravenously administered cholesterol-4-^{14}C-labeled lipoprotein.

We have recently had an opportunity to study the responses of four different species of monkeys to a cholesterol-containing diet. All are male animals and consume either a control diet containing 25 percent

TABLE 2 Total Serum Cholesterol Levels of Cholesterol-Fed Female Squirrel Monkeys before and during the Breeding Season

Group	No.	Non-Breeding Season				Breeding Season		
		12-27-67	1-27-68	2-28-68	3-27-68	4-30-68	5-28-68	6-26-68
Safflower Oil	5	412 ± 53[a]	372 ± 18	442 ± 46	305 ± 27	244 ± 50	200 ± 21	213 ± 25
Lard	4	376 ± 111	503 ± 175	402 ± 148	295 ± 70	243 ± 24	236 ± 23	250 ± 29
Butter	7	467 ± 37	559 ± 36	471 ± 51	329 ± 13	206 ± 13	182 ± 16	221 ± 24

[a]Serum cholesterol levels are expressed as the mean followed by the standard error of the mean.

lard and no additional cholesterol, or a diet containing 25 percent lard and 1 mg per calorie of cholesterol. Table 3 is a summary of the response of these monkeys to the diets. It can be seen that major differences exist among the animals in the level of hypercholesterolemia that result from cholesterol feeding. We feel that essential to further elucidation of primate models is a better understanding of the mechanisms that account for these differences in serum cholesterol levels.

Pigs have received their share of attention in the study of atherosclerosis. It has been said that in general dietary habits pigs are more like human beings than any other animal. Pigs fed increased amounts of cholesterol develop alterations in their serum lipids and blood-clotting mechanisms, and increased severity of atherosclerotic lesions.[29] Most of the studies have been on young animals, and conducted over relatively short periods of time; hence the full potential usefulness of these animals has yet to be determined.

One of the most promising possibilities for their usefulness is for the study of cerebral vascular disease. Considerable cerebral artery atherosclerosis and associated brain lesions have been reported in old pigs that have been fed garbage.[13]

Dogs, with which most surgeons and physiologists feel at home, have had some small place in experimental atherosclerosis research. The results of most studies have suggested that dogs do not develop atherosclerosis unless thyroid activity is suppressed and large amounts of cholesterol are added to the diet.[36] With such a regimen, serum lipids rise and lipids appear first in the media of small arteries, later in the aortic media. Intimal lesions appear later. The eventual effects of this lipid-containing arterial lesion are not known; long-term dog experi-

TABLE 3   The Effect of a Cholesterol-Containing Diet on the Serum Cholesterol Levels of *Macaca arctoides, Lagothrix lagothricha, Cercopithecus aethiops,* and *Saimiri sciureus*[a,b]

	N	0.5% Cholesterol Diet	N	Control Diet
Macaca arctoides	5	735 ± 29.3	5	151 ± 4.9
Cercopithecus aethiops	4	464 ± 13.9	5	158 ± 5.1
Lagothrix lagothricha	3	127 ± 6.2	3	103 ± 3.5
Saimiri sciureus	10	431 ± 20.9	10	159 ± 4.4

[a]Mean value of 11 samplings for the group over a one-year period ± standard error of the mean.
[b]Adapted from Lehner *et al.* Proceedings of the Second Conference on Experimental Medicine and Surgery in Primates. In press.

TABLE 4   Effect of Diet on Atherosclerosis among Three Breeds of Pigeons

Breed	Sex	No. Examined	Aortic Atherosclerosis[a]	Coronary Lesion Prevalence[b]	Coronary Lesion Extent[c]	Coronary Artery Stenosis[d]	Total Serum Cholesterol[e]
White Carneau	Male	7	42 ± 13	26 ± 6	38 ± 8	86 ± 10	$1,940 \pm 379$
	Female	13	31 ± 6	24 ± 6	34 ± 6	80 ± 9	$1,414 \pm 223$
Show Racer	Male	9	4 ± 1	44 ± 10	40 ± 9	89 ± 10	$1,932 \pm 178$
	Female	11	6 ± 1	26 ± 6	11 ± 3	49 ± 13	$1,212 \pm 198$
F_2	Male	7	12 ± 3	50 ± 6	47 ± 7	99 ± 1	$2,190 \pm 276$
	Female	10	13 ± 4	10 ± 3	10 ± 4	36 ± 12	$1,297 \pm 206$

[a] Expressed as the mean percent intimal surface of aorta involved with plaques followed by the standard error of the mean.

[b] Expressed as the mean percent of coronary arteries with plaques in 15 step sections of left ventricular and septal myocardium followed by the standard error of the mean.

[c] Expressed as the mean percent of lumen obliterated by plaque among those coronary arteries with lesions followed by the standard error of the mean.

[d] Expressed as the mean percent of lumen obliterated by the most stenotic lesion found in each bird of the groups followed by the standard error of the mean.

[e] Expressed as the mean total serum cholesterol determined at the termination of the experiment followed by the standard error of the mean.

ments have not been done. More recently atherosclerosis has been induced in dogs by feeding them a semisynthetic diet that was thiouracil-free and contained cholesterol and hydrogenated coconut oil.[24]

We have studied the potential usefulness of domestic cats as animal models of atherosclerosis.[26] For 12 months test cats were fed a high-fat, cholesterol-containing diet; control cats were fed the same diet without added cholesterol. Test cats did not develop sustained hyper-cholesterolemia when the diet contained 0.5 percent cholesterol; however, marked hypercholesterolemia occurred when the added cholesterol content was increased to 2.0 percent. At necropsy, all five test cats had aortic atherosclerosis, manifested as fatty streaks and elevated plaques. Two test cats had coronary artery atherosclerosis of the proximal main branches of the right or left coronary arteries. No atherosclerosis was seen in control cats. The anatomic location of the coronary artery lesions is a possible advantage of this model.

Pigeons have appeared relatively recently on the research scene, initially attracting attention because of the marked resemblance of their naturally-occurring atherosclerosis to that in man, and because of considerable variation in the pattern of atherosclerosis in different breeds.[7] When relatively small amounts of cholesterol are fed to pigeons their serum lipids rise, and there is marked aggravation of their arterial lesions.[10] Myocardial infarction occurs from obstructive coronary lesions.[33] As in the rabbit, however, cholesterol feeding also affects other organs. The pigeon studies made thus far have shown no beneficial effects of many pharmacologic agents, have demonstrated that the interrelationships of various dietary factors are important in assessing atherogenic diets, have provided an opportunity to develop special substrains, and have given us material for studies in arterial metabolism.[20]

For some time there has been controversy about the mechanisms of "female protection" among the females of some populations of human beings. The controversy stems from the observations that the females of some populations seem somewhat protected against coronary disease while those of others do not. Among several strains of pigeons there are ideal models for studying this phenomenon. Female Show Racer pigeons, when fed cholesterol, have strikingly less coronary artery atherosclerosis than do male birds of that breed; in contrast, no sex difference exists among male and female pigeons of the White Carneau breed fed the same diet for the same length of time (Table 4). Birds produced by a cross of the Show Racer and White Carneau pigeons seem intermediate in their susceptibility to atherosclerosis, while they tend to resemble their Show Racer parents in their susceptibility to coronary artery disease.

Pigeons have another unique advantage over other animal models in that lesions develop in predictable areas of the aorta after short-term cholesterol feeding. Consequently, this species is particularly valuable in studies on the early metabolic changes associated with lesion development.

Turkeys are so liable to develop dissecting aneurysms of the aorta that it is a major economic problem of turkey growers.[14] The dissection may start in areas weakened by atherosclerosis, for turkeys do develop atherosclerosis that can be aggravated by cholesterol feeding. Turkeys also have arterial blood pressure levels much higher than those of most other birds, and reduction of blood pressure by means of reserpine has lowered the incidence of aortic rupture. The use of the turkey in pure atherosclerosis research work would seem to have limited promise, since these birds offer no advantage over pigeons in terms of suitability for surgical procedures, their cost is higher, and they present greater husbandry difficulties.

Some experimental work in atherosclerosis has involved other than intact animals as the model system. Tissue cultures of various components of artery walls, and of intact segments of arteries, have been made. The arteries in question have been from human and from nonhuman sources. Various aspects of intermediary and lipid metabolism have been studied. One would hope in these studies, by examining metabolic activity in arteries prone to disease and in others less susceptible, to be able to detect qualitative or quantitative differences. Attempts could then be made to correct the metabolic abnormality by a variety of means.

Our research group has been studying both animal and human arteries by a perfusion technique.[20] The use of this model system has enabled us to characterize the changes in lipid metabolism of the arterial wall associated with plaque development.

SUMMARY

In this review we summarized our experiences with, and our thoughts about, some of the more commonly used animal models in atherosclerosis research. Certain atherosclerosis characteristics of rabbits, chickens, rats, nonhuman primates, dogs, cats, and turkeys have been presented.

Supported in part by grants from the National Institutes of Health (HE 04722, HE 04352, HE 04371, and FR 00180).

REFERENCES

1. Adel, H. N., Q. B. Deming, and L. Brun. 1969. Genetic hypercholesterolemia in rats. Suppl. III. Circulation 39 and 40:1.

2. Andrus, S. B., O. W. Portman, and A. J. Riopelle. 1968. Comparative studies of spontaneous and experimental atherosclerosis in primates. II. Lesions in chimpanzees including myocardial infarction and cerebral aneurysms. Progr. Biochem. Pharmacol. 4:393–419.

3. Anitschkow, N. 1914. Über die Atherosclerose der Aorta beim Kaninchen und über derin Entstehungsbedingungen. Beitr. Path. Anat. 59:308–348.

4. Bullock, B. C. 1967. Effect of age and diet on coronary artery atherosclerosis in *Cebus albifrons*. Federation Proc. 26:371.

5. Bullock, B. C., T. B. Clarkson, N. D. M. Lehner, H. B. Lofland, Jr., and R. W. St. Clair. 1969. Atherosclerosis in *Cebus albifrons* monkeys. Exptl. Molec. Path. 10:39–60.

6. Chawla, K. K., C. D. S. Murthy, R. N. Chakravarti, and P. N. Chhuttani. 1967. Arteriosclerosis and thrombosis in wild rhesus monkeys. Am. Heart Journal 73:85–91.

7. Clarkson, T. B., R. W. Prichard, M. G. Netsky, and H. B. Lofland. 1959. Atherosclerosis in pigeons. Its spontaneous occurrence and resemblance in human atherosclerosis. Arch. Path. 68:143–147.

8. Clarkson, T. B., B. C. Bullock, and N. D. M. Lehner. 1968. Pathologic characteristics of atherosclerosis in New World monkeys. Progr. Biochem. Pharmacol. 4:420–428.

9. Clarkson, T. B., H. B. Lofland, B. C. Bullock, N. D. M. Lehner, R. W. St. Clair, and R. W. Prichard. 1969. Atherosclerosis in some species of New World monkeys. Ann. N. Y. Acad. Sci. 162:103–109.

10. Clarkson, T. B., and H. B. Lofland. 1961. Effects of cholesterol-fat diets on pigeons susceptible and resistant to atherosclerosis. Circ. Res. 9:106–109.

11. Constantinides, P. 1965. Experimental atherosclerosis in the rabbit. *In* J. C. Roberts and R. Straus (eds.) Comparative Atherosclerosis. p. 276–290. Harper and Row, New York.

12. Dauber, D. V. 1944. Spontaneous arteriosclerosis in chickens. Arch. Path. 88:46.

13. Detweiler, D. K., H. L. Ratcliffe, and H. Luginbahl. 1968. The significance of naturally occurring coronary and cerebral arterial disease in animals. Ann. N.Y. Acad. Sci. 149:868–881.

14. Durrell, W. B., B. S. Pomeroy, W. S. Carr, and A. C. Jerstad. 1952. *In* Proceedings of the 89th Annual Meeting of the American Veterinary Medical Association. p. 280–281.

15. Finlayson, R. 1965. Spontaneous arterial disease in exotic animals. J. Zool. 147:239–343.

16. Howard, A. N., G. A. Gresham, D. E. Bowyer, and F. T. Lindgren. 1968. Aortic and coronary atherosclerosis in baboons. Progr. Biochem. Pharmacol. 4:438–444.

17. Katz, L. N., and J. Stamler. 1953. Experimental atherosclerosis. C. C Thomas, Springfield, Ill. p. 375.

18. Kramsch, D. M., and W. Hollander. 1968. Occlusive atherosclerotic disease of the coronary arteries in monkey (*Macaca irus*) induced by diet. Exptl. Molec. Path. 9:1–22.

19. Lindsay, S., and I. L. Chaikoff. 1966. Naturally occurring arteriosclerosis in nonhuman primates. J. Atheroscl. Res. 6:36–61.

20. Lofland, H. B. 1968. Recent advances in arterial metabolism: The whole artery. Prog. Biochem. Pharmacol. 4:211–217.

21. Lofland, H. B., R. W. St. Clair, J. E. MacNintch, and R. W. Prichard. 1967. Atherosclerosis in New World primates. Biochemical studies. Arch. Path. 83:211–214.

22. Malinow, M. R., and C. A. Maruffo. 1965. Aortic atherosclerosis in free-ranging howler monkeys (*Alouatta caraya*). Nature 206:948.

23. Malinow, M. R., C. A. Maruffo, and A. M. Perley. 1966. Experimental atherosclerosis in squirrel monkeys (*Saimiri sciurea*). J. Path. Bacteriol. 92:491–510.

24. Malmros, H., and N. H. Sternby. 1968. Induction of atherosclerosis in dogs by a thiouracil-free semisynthetic diet containing cholesterol and hydrogenated coconut oil. Prog. Biochem. Pharmacol. 4:482–487.

25. Mann, G. V., S. B. Andrus, A. McNally, and F. J. Stare. 1953. Experimental atherosclerosis in cebus monkeys. J. Exptl. Med. 98:195–218.

26. Manning, P. J., and T. B. Clarkson. In press. Diet-induced atherosclerosis of the cat. Arch. Path.

27. McGill, H. C., Jr., J. P. Strong, R. L. Holman, and N. T. Werthesson. 1960. Arterial lesions in the Kenya baboon. Circ. Res. 8:670–679.

28. Middleton, C. C., T. B. Clarkson, H. B. Lofland, and R. W. Prichard. 1967. Diet and atherosclerosis of squirrel monkeys. Arch. Path. 83:145–153.

29. Moreland, A. F., T. B. Clarkson, and H. B. Lofland. 1963. Atherosclerosis in "miniature" swine. Arch. Path. 76:203–210.

30. Paterson, J. C., S. J. Slinger, and K. M. Gartley. 1948. Experimental coronary sclerosis. Arch. Path. 45:306.

31. Pick, R., J. Stamler, and L. Katz. 1952. The inhibition of coronary atherosclerosis by estrogen in cholesterol-fed chicks. Circulation 6:276–280.

32. Portman, O. W., and S. B. Andrus. 1965. Comparative evaluation of three species of New World monkeys for studies of dietary factors, tissue lipids and atherogenesis. J. Nutr. 87:429–438.

33. Prichard, R. W., T. B. Clarkson, and H. B. Lofland. 1963. Myocardial infarcts in pigeons. Am. J. Path. 43:651–659.

34. Prior, J. T., and D. D. Ziegler. 1965. Regression of experimental atherosclerosis. Arch. Path. 80:50–57.

35. Prior, J. T., D. M. Kurtz, and D. D. Ziegler. 1961. The hypercholesterolemic rabbit. Arch. Path. 71:672–684.

36. Schenk, E. A., I. Penn, and S. Schwartz. 1965. Experimental atherosclerosis in the dog. Arch. Path. 80:102–109.

37. Scott, R. F., E. S. Morrison, J. Jarmolych, S. C. Nam, M. Kroms, and F. Coulston. 1967a. Experimental atherosclerosis in rhesus monkeys. I. Gross and light microscopy features and lipid values in serum and aorta. Exptl. Molec. Path. 7:11–33.

38. Scott, R. F., R. Jones, A. S. Daoub, O. Zumbo, F. Coulston, and W. A. Thomas. 1967b. Experimental atherosclerosis in rhesus monkeys. II. Cellular elements of proliferative lesions and possible role of cytoplasmic degeneration in pathogenesis as studied by electron microscopy. Exptl. Molec. Path. 7:34–57.
39. Strong, J. P. 1965. Arterial lesions in primates. *In* J. C. Roberts and R. Straus Comparative Atherosclerosis. Harper and Row. New York. p. 244–252.
40. Strong, J. P., and H. C. McGill. 1967. Diet and experimental atherosclerosis in baboons. Am. J. Path. 50:669–690.
41. Taylor, C. B. 1965. Experimentally induced arteriosclerosis in nonhuman primates. *In* Comparative Atherosclerosis. J. C. Roberts and R. Straus, eds., Harper (Hoeber), New York.
42. Taylor, C. B., Pacita Manalo-Estrella, and G. E. Cox. 1963a. Atherosclerosis in rhesus monkey. V. Marked diet-induced hypercholesteremia with xanthomatosis and severe atherosclerosis. Arch. Path. 76:239–249.
43. Taylor, C. B., D. Patton, and G. E. Cox. 1963. Atherosclerosis in rhesus monkeys. VI. Fatal myocardial infarction in a monkey fed fat and cholesterol. Arch. Path. 76:404–412.
44. Thomas, W. A., W. S. Hartroft, and R. M. O'Neal. 1959. Modification of diets responsible for induction of coronary thromboses and myocardial infarcts in rats. J. Nutr. 69:325–337.
45. Wissler, R. W., L. E. Frazier, R. H. Hughes, and R. A. Rasmussen. 1962. Atherogenesis in the cebus monkey. I. A comparison of three food fats under controlled dietary conditions. Arch. Path. 74:312–322.
46. Wissler, R. W., G. S. Getz, D. Vesselinovitch, L. E. Frazier, and R. H. Hughes. 1966. Acute severe experimental atherosclerosis in rhesus monkeys. Fed. Proc. 25:597.
47. Wissler, R. W., D. Vesselinovitch, G. S. Getz, and R. H. Hughes. 1967. Aortic lesions and blood lipids in rhesus monkeys fed three food fats. Fed. Proc. 26:371.

Animal Models in
Hemophilia Research: A Review[*]

MERLE E. MUHRER

For many years, hemophilia has been used to describe any abnormal tendency toward bleeding. In recent times, the word *hemophilia* has been restricted to describing a specific abnormality characterized by a deficiency of one of the coagulation factors called the antihemophilic factor (AHF) and more recently renamed factor VIII by the International Nomenclature Committee.[43] Hemophilia poses a most serious social problem for those unfortunate human beings suffering from it and causes fear, pain, crippling, and death. It is possible that experimental work could be done on hemophilic animal models that might provide information to assist both man and animals afflicted with a hemophilia-like condition.

CHOICE OF THE ANIMAL MODEL

Hypotheses about disease processes in man often cannot be tested directly. Therefore, experimental medicine depends heavily on animal models.[15, 21] Animal models mean different things to different people. The classic example of an animal model is the guinea pig. To the vitamin nutritionist, this may be the model for vitamin C and scurvy experimentation; for the pathologist, the guinea pig may be a test animal for testing drugs to be used against tuberculosis. For the endocrinologist, this same animal may be used for estimation of hormone balance.

*Contribution from the Missouri Agricultural Experiment Station Journal Series, No. 5743.

The guinea pig is also valuable as an animal model in hemostasis research because of capillary hemorrhage following collagen depletion during ascorbic acid deficiency.

The mark of a good animal model is predictability. The physician should be able to predict from the responses and results of the model animal what the responses and results will be in corresponding human cases.

THE DOG AS A HEMOPHILIA MODEL

An inherited hemostatic defect considered nearly identical to human hemophilia is more frequently reported in dogs than in any other animal. This inherited disorder of blood coagulation is characterized by hemorrhage following minor trauma, a prolonged coagulation time, and a sex-linked recessive pattern of genetic transmission.[7, 9, 16] A blood plasma deficiency of factor VIII is generally accepted as the major defect in the hemophilic dog.[10] The inheritance of factor VIII deficiency in canines follows a sex-linked recessive pattern.[37] Most of the deficient puppies die early in life if they are not treated. The usual treatment is normal whole blood or plasma transfusions. The factor VIII deficiency is usually due to lack of production of the factor rather than to an inhibitor or excess utilization.[28] The deficiency of factor VIII causes the dogs to develop hemarthrosis, multiple hematomas, and apparently spontaneous hemorrhages into most body parts, including muscles, joints, and soft tissues. In addition, there is excessive bleeding following injury, dental extractions, and surgery. The factor VIII deficiency limits activation of factor X and results in a slow conversion of prothrombin to thrombin due to a lack of sufficient intrinsic activated thromboplastin.

Hemophilia-like inherited hemostatic defects are known to occur in a number of breeds of dogs. Hemophilia has been reported in Irish setters,[14, 16] German shepherds,[36] collies,[36] Labrador retrievers,[1] beagles,[9] Shetland sheep dogs,[44] and greyhounds,[41] as well as in other breeds.

A hemorrhagic defect was found in three male Weimaraner dogs from a single litter and in a male Chihuahua.[23] The clinical signs and laboratory findings in these dogs were consistent with a diagnosis of hemophilia.

Buckner *et al.*[10] presented laboratory and clinical evidence of canine hemophilia in the Vizsla breed of dogs. These animals were successfully treated with frozen normal canine plasma. Dogs with congenital defects of factor VII, VIII, or IX were studied by Honig et al.[18] It was concluded that the initial platelet interaction with the vessel wall and surrounding tissues is not dependent on coagulation. However, an adequate and intact intrinsic pathway of coagulation is necessary for stabilization of the hemostatic plug after it is formed. Many other examples could be given of the use of canine hemophilia animal models in the elucidation of desirable hemostatic principles and practices.

THE HORSE AS A HEMOPHILIA MODEL

The first case of hemophilia in horses was probably reported by Archer in 1961.[2] A year later, Nossell et al. reported another case of equine hemophilia.[34] The hemostasis tests indicated a hemophilia picture but the genetic pattern was not consistent with this diagnosis. The most convincing case was reported by Sanger et al.[39] The hemorrhagic disorder in a Standardbred foal was indistinguishable from classical hemophilia. Three other male foals from the same dam died from hemorrhage; the mare colts were normal. These foals developed hematocysts sporadically on the body and limbs. Blood transfusions of normal horse blood were given once or twice weekly to keep the foals alive. The animals were hemorrhagic at birth. A later case of equine hemophilia has been reported by Australian workers, Hutchins et al.[19]

According to Bell et al.,[4] it appears that a mild deficiency of factor VIII is the normal state in the horse and is present in both sexes. From the level of the clotting time and the thromboplastin generation test, the horses studied would correspond to only moderately severe human hemophiliacs. Because of the rarity of spontaneous hemorrhage in the horse, there must be a compensatory mechanism to account for the few cases of pathological hemorrhage. A possible explanation may be an increase in extrinsic or tissue thromboplastin activity. These findings may prove to be a useful tool for the investigation of hemophilia in other species, including man.

The horses studied by Bell et al.[4] had a prolonged coagulation time and poor clot retraction even with normal platelet count. These defects could be corrected by giving normal human plasma. This is further evidence of the value of the horse as a model to study human hemostasis.

CATTLE USED IN HEMOSTASIS RESEARCH

Schofield discovered a bleeding condition in cattle due to eating spoiled sweet clover hay.[40] His work led to the isolation of dicumarol by Link[27] and the relation of this oral anticoagulant to the prothrombin complex. Before the relations of dicumarol, prothrombin, and vitamin K were established, many experimental animal models were used, including dogs, rabbits, and chickens. Although these are not hemophilia models, there are close hemostasis relationships. Muhrer and Gentry[32] used rabbits for animal models to study a hemorrhagic factor in moldy lespedeza hay that was found to be a dicumarol-like anticoagulant causing undesirable hemorrhage in cattle.

Dicumarol and related anticoagulant compounds have been used extensively in human medicine to prevent intravascular coagulation and thrombosis. In the absence of excessive levels of vitamin K, liver synthesis of prothrombin and accessory factors is greatly curtailed by dicumarol-like compounds. Coagulability of blood is then decreased along with the danger of thrombosis, although the danger of hemorrhage is increased. The discoveries made in a cattle-model bleeding syndrome have greatly enhanced human medicine and have provided a treatment for one of our major causes of death.

SWINE AS HEMOPHILIA MODELS

Through considerable effort, vigilance, and interdisciplinary cooperation, researchers at the University of Missouri have been able to maintain a herd of swine afflicted with an inherited hemorrhagic disorder. No other domestic or wild animal in the world has been reported to have this same hemophilia-like disorder. These swine are unique animal disease models used for hemostasis research.

One of the first research reports concerning an animal model of an inherited human-like bleeding disease came from the University of Missouri. In 1941, Hogan, Muhrer, and Bogart first reported a hemorrhagic tendency in a herd of inbred swine.[17] They observed that the bleeding time of the affected animals was normal if determined by Duke's method, but if an incision of any extent was made, the bleeding time was prolonged indefinitely. They also reported that blood from the bleeder swine is normal in quantity of prothrombin, serum calcium, and fibrinogen. Mertz confirmed the finding of a normal

Duke's bleeding time.[30] However, by the use of the Copley and Lalich saline bleeding time method,[11] he was able to show a gross abnormality in the bleeder swine. This saline bleeding time method is presently being used as one of the major means of classifying the swine into bleeder and nonbleeder groups. A study of the inheritance by Bogart and Muhrer[5] indicated that the trait was inherited as an autosomal recessive characteristic and that one or more modifying genes determined the severity of the abnormality.

Further characterization of the disorder was achieved by Brinkhous *et al.*[8] They found the plasma antihemophilic factor (factor VIII) to be greatly reduced in the affected animals. They suggested that the clotting abnormality was similar to hemophilia seen in man and dog.

Cornell and Muhrer performed numerous coagulation tests on the swine as well as a specific assay for factor VIII.[12] They confirmed that the concentrations of prothrombin, fibrinogen, platelets, and calcium were adequate for normal coagulation. Also, they reported that the clotting abnormality was not due to an inhibitor of thromboplastin or factor IX deficiency. Using a specific factor VIII assay the bleeder swine were found to be grossly deficient in this essential coagulation system protein.

Working with Dr. Brinkhous and staff, Muhrer *et al.* reported that infusions of normal plasma or serum resulted in an unusual response of plasma factor VIII levels.[33] The initial rise in factor VIII following plasma infusion was maintained with some fluctuation for 24 hours after the infusion. These findings are in contrast with infusion experiments with classical hemophiliacs. In the latter disease, after plasma infusions, factor VIII levels rise initially but then steadily decline; serum infusions do not immediately or directly increase factor VIII levels.

Bleeder swine with a low amount of factor VIII were infused with normal swine plasma. The factor VIII level went up as would be expected. A decrease in factor VIII succeeding the initial increase after the dilution effect of infused factor VIII was no longer evident. Serum with very little factor VIII was then infused and again the factor VIII was increased in a delayed fashion. Factor VIII levels encountered after the infusion of either plasma or serum increased to levels above the amount accounted for coming from the infused materials. These results were evidence for a factor VIII-stimulating factor that has been named "factor S."

Recently, Bowie *et al.* isolated a fraction from either plasma or serum which, when infused into the bleeder swine, will stimulate a rise in factor VIII levels as do plasma or serum infusions.[6] The factor S fraction appears to be unrelated to fibrinogen or to factor VIII.

Twenty-five infusions of normal pig plasma, serum, or their sephadex G-200 and DEAE-cellulose chromatographic fractions were made in pigs with the bleeding diathesis. The infusion results were characterized by increases in factor VIII activity significantly above that actually infused. When plasma was fractionated, fraction I from sephadex G-200 columns contained fibrinogen, factor VIII, and factor VIII-stimulating factor (factor S). When serum was fractionated, the stimulating factor was in this zone, but fibrinogen and demonstrable factor VIII were lacking. The molecular weight of the stimulating factor may exceed 200,000. When plasma was fractionated, the last protein fraction eluted from the DEAE-cellulose columns contained factor VIII and the stimulating factor. When serum was fractionated, only the stimulating factor was found in this zone. Both factors were clearly separate from fibrinogen, which emerged in the first protein zone. Since the stimulating factor was eluted from DEAE-cellulose at high ionic strength (concentrated in the last major elution peak), the factor S molecule seemed to be highly charged at pH 7.4.

Evidence is accumulating that there is a factor that controls the production or release of a key material in the hemostatic mechanism. Further understanding and elucidation of this factor and the mechanisms involved is highly important to advancement in scientific medicine.

Studies on the platelets from the bleeder swine began when Muhrer et al. measured platelet fragility by passing platelet-rich plasma through different concentrations of sodium chloride.[31] Fibrin precipitation times were determined on the diluted activated plasma after recalcification. The results led to the conclusion that platelets from bleeder swine liberated less thromboplastin (called platelet factor 3) and therefore were more stable than those of normal swine. It was assumed that the conditions in the normal and bleeder plasmas were the same and that differences in the results were due to differences in the platelets. It has since been discovered that the bleeder swine are grossly deficient in factor VIII, which is necessary for a normal clotting time and perhaps led to the differences observed. Recently, we demonstrated that bleeder swine have significantly less platelet adhesion than do normal swine as measured by the Salzman[38] filter method. Because of the controversy over the validity of this method, Cornell et al. also studied platelet adhesion in these animals.[13] After developing a method subject to fewer variables, they were able to confirm reduced adhesion in bleeder swine platelets.

Platelet morphology of the bleeder swine was studied in whole mounts with the electron microscope by Kahn *et al.*[22] The bleeder swine platelets retain their granulomere significantly more often than normal in the later stages of metamorphosis. This difference does not seem to be related to the levels of factor VIII.

Three separate though often superimposed defects are found in these animals: the plasma from the bleeder animals is deficient in factor VIII, the platelets are abnormal, and the bleeding pattern indicates a vascular defect. Perhaps hemorrhage in man is often due to more than one defective hemostatic mechanism, which may constitute a safety measure that we have not fully appreciated. A study of the combinations of defects in hemostasis may make the Missouri bleeder swine a unique and most valuable model for research on human diseases involving hemostasis.

HEMOPHILIA VERSUS THROMBOPHILIA

Thrombophilia, the affinity for thrombosis, may be more important in the health of man and animals than hemophilia, the affinity for hemorrhage. Hypercoagulability and thrombosis probably cause more deaths than hypocoagulability and hemorrhage. Thrombosis is often associated with serious threats to life in atherosclerotic heart disease, stroke, severe tissue injury, and many other causes of intravascular coagulation. Perhaps the most important balance between life and death is closely related to the balance between hemorrhage and thrombosis. The balance is often shifted from one extreme to the other. Non-fatal intravascular clotting may consume a sufficient amount of the coagulation factors to produce a fatal bleeding syndrome in the victim. This phenomenon has been described by Owen *et al.*[35] as the Intravascular-Coagulation-Fibrinolysis (ICF) syndrome. Perhaps a more descriptive label would be the ICF-H syndrome, referring to the Intravascular-Coagulation-Fibrinolysis-Hemorrhage sequence of events that often causes a critical threat to life. This sequence may explain the real cause of death in a number of diseases. Animal models are important for the study of this newly-defined cause of death.

SUMMARY

The study of diseases in animals has made a profound contribution to the understanding of diseases in man. The balance between life and death in many diseases is often a balance in hemostasis between thrombosis from hypercoagulability and hemorrhage from hypocoagulability. Animal models for the study of hemophilia-like diseases are described and include dogs, horses, cattle, and swine. The Missouri hemophilia-like swine are unique and valuable as models for hemostasis research because of a triple defect in factor VIII, platelets, and vessels.

ACKNOWLEDGMENTS

The Missouri research was supported by a National Institutes of Health Grant, HE 07181. Cooperation by Dr. K. M. Brinkhous and staff at the University of North Carolina on comparative hemophilia studies is recognized and appreciated. Cooperation by Dr. C. A. Owen and staff at the Mayo Clinic on characterization and testing of factor S is recognized and appreciated. Housing for the hemophilia-like swine is presently provided by the Sinclair Comparative Medicine Research Farm. Apologies are hereby made for references omitted and assistance unrecognized.

REFERENCES

1. Archer, R. K., and R. S. T. Bowden. 1959. A case of true hemophilia in a Labrador dog. Vet. Rec. 71:560–561.
2. Archer, R. K. 1961. True haemophilia (hemophilia A) in a thoroughbred foal. Vet. Rec. 73:338–340.
3. Beecher, H. K. 1969. Human studies. Science 164:1256–1258.
4. Bell, W. N., S. C. Tomlin, and R. K. Archer. 1955. The coagulation mechanism of the blood of the horse with particular reference to its haemophiloid status. J. Comp. Path. and Therap. 65:255–261.
5. Bogart, R., and M. E. Muhrer. 1942. The inheritance of a hemophilia-like condition in swine. J. Hered. 33:59–64.
6. Bowie, E. J. W., J. Y. S. Chan, P. Didisheim, J. H. Thompson, Jr., M. E. Muhrer, and C. A. Owen. 1967. The factor VIII "stimulating factor" in porcine von Willebrand's disease. J. Lab. Clin. Med. 70:991.
7. Brinkhous, K. M., and J. B. Graham. 1950. Hemophilia in the female dog. Science 111:723–724.
8. Brinkhous, K. M., F. C. Morrison, Jr., and M. E. Muhrer. 1952. Comparative study of clotting defects in human, canine, and porcine hemophilia. Fed. Proc. 11:409.
9. Brock, W. E., R. G. Buckner, J. W. Hampton, R. M. Bird, and C. E. Wulz. 1963. Canine hemophilia. Arch. Path. 76:464–469.
10. Buckner, R. G., J. M. Hamptom, R. M. Bird, and W. E. Brock. 1967. Hemophilia in the Vizsla. J. Small Animal Practices 8:511–519.

11. Copley, A. L., and J. J. Lalich. 1941. Bleeding time in men. Am. J. Physiol. 133:246–247.

12. Cornell, C. N., and M. E. Muhrer. 1964. Coagulation factors in normal and hemophilia-type swine. Am. J. Physiol. 206:926–928.

13. Cornell, C. N., R. G. Cooper, R. A. Kahn, and S. Garb. 1969. Platelet adhesiveness in normal and bleeder swine as measured in a celite system. Am. J. Physiol. 216:1170–1175.

14. Field, R. A., C. G. Richard, and S. B. Hutt. 1946. Hemophilia in a family of dogs. Cornell Vet. 36:293–299.

15. Frenkel, J. K. 1969. Choice of animal models for the study of disease processes in man. Fed. Proc. 28:160–161.

16. Graham, J. B., J. A. Buckwalter, L. J. Hartley, and K. M. Brinkhous. 1949. Canine hemophilia; observations on the course, the clotting anomaly, and the effect of blood transfusions. J. Exp. Med. 90:97–111.

17. Hogan, A. G., M. E. Muhrer, and R. Bogart. 1941. A hemophilia-like disease in swine. Proc. Soc. Exp. Biol. Med. 48:217–219.

18. Honig, T., H. C. Rowsell, W. J. Dodds, L. Jorgensen, and J. F. Mustard. 1967. Experimental hemostasis in normal dogs and dogs with congenital disorders of blood coagulation. Blood 30:636–668.

19. Hutchins, D. R., E. E. Lepherd, and I. G. Crook. 1967. A case of equine hemophilia. Aust. Vet. J. 43:83–87.

20. Ingelfinger, F. J. 1968. Royal blights. New Eng. J. Med. 278:506–507.

21. Jones, T. C. 1969. Mammalian and avian models of diseases in man. Fed. Proc. 28:162–164.

22. Kahn, R. A., R. G. Cooper, C. N. Cornell, and M. E. Muhrer. 1967. Platlet morphology in Missouri bleeder swine. Trans. Mo. Acad. Sci. 1:86.

23. Kaneko, J. J., D. R. Cordy, and G. Carlson. 1967. Canine hemophilia resembling classic hemophilia A. J. Am. Vet. Med. Ass. 150:15–21.

24. Kitchen, H. 1968. Comparative biology: Animal models of human hematologic diseases. Pediat. Res. 2:215–229.

25. Ladimer, J. 1967. Rights, responsibilities, and protection of patients in human studies. J. Clin. Pharmacol. 7:125–130.

26. Leader, R. W. 1967. The kinship of animal and human diseases. Scientific American 216:110–116.

27. Link, K. P. 1943. The anticoagulant from spoiled sweet clover hay. The Harvey Lectures 39:162–216.

28. McLester, W. D., and R. H. Wagner. 1965. Antibody to antihemophilia factor and its lack of reaction with hemophilia plasma. Am. J. Physiol. 208:499.

29. Massie, R. K. 1967. Nicholas and Alexandria. Atheneum Publication, New York.

30. Mertz, E. T. 1942. The anomaly of a normal Duke's and a very prolonged saline bleeding time in swine suffering from an inherited bleeding disease. Am. J. Physiol. 136:360–362.

31. Muhrer, M. E., R. Bogart, and A. C. Hogan. 1944. Estimate of platelet fragility. Am. J. Physiol. 144:449–453.

32. Muhrer, M. E., and R. F. Gentry. 1948. A hemorrhagic factor in moldy lespedeza hay. Mo. Expt. Sta. Res. Bul. 429:1–11.

33. Muhrer, M. E., E. Lechler, C. N. Cornell, and J. L. Kirkland. 1965. Antihemophilic factor levels in bleeder swine following infusions of plasma or serum. Am. J. Physiol. 208:508–510.
34. Nossel, H. L., R. K. Archer, and R. G. MacFarlane. 1962. Equine haemophilia: Report of a case and its response to multiple infusions of heterospecific AHG. Brit. J. Haemat. 8:335.
35. Owen, C. A., H. C. Oels, E. J. W. Bowie, P. Didisheim, and J. H. Thompson. 1969. Chronic intravascular coagulation fibrinolysis syndrome. Wayne State University Annual Symposium on Blood 17:22.
36. Rowsell, H. C. 1963. Hemorrhagic disorders in dogs: Their recognition, treatment and importance. 12th Gaines Veterinary Symposium, 9–16.
37. Rowsell, H. C., and J. F. Mustard. 1966. Hemostasis in domestic animals. Can. J. Med. Technol. 28:2.
38. Salzman, E. W. 1963. Measurement of platelet adhesiveness. J. Lab. Clin. Med. 62:724–735.
39. Sanger, V. L., R. E. Mairs, and A. L. Trapp. 1964. Hemophilia in a foal. J. Am. Vet. Med. Ass. 144:259–264.
40. Schofield, F. W. 1922. A brief account of a disease in cattle simulating hemorrhagic septicaemia due to feeding sweet clover. Can. Vet. Rec. 3:74.
41. Sharp, A. A., and G. W. R. Dike. 1963. Hemophilia in the dog—Treatment with heterologous anti-haemophilic globulin. Thromb. et Diathes. Haemorrhagica 10:494–501.
42. Wolfle, D. 1960. Research with human subjects. Science 132:989.
43. Wright, I. S. 1959. Nomenclature of blood clotting factors. J.A.M.A. 170:325–328.
44. Wurzel, H. A., and W. C. Lawrence. 1961. Canine hemophilia. Thromb. et Diathes. Haemorrhagica 6:98–103.

Animal Models for the Study
of Hemoglobinopathies

HYRAM KITCHEN, ETHEL OLIVER KOUBA,
and CAROLINE W. EASLEY

The hemoglobinopathies of man have deleterious effects on individuals. Severe anemia is often accompanied by gross pathological changes. In contrast, few, if any, harmful hemoglobinopathies have been documented in other animal species.[1, 2] However, several naturally occurring biological phenomena related to animal hemoglobins show variable expressions of quantitative control of hemoglobin production and may serve as ideal models for the study of mechanisms of the human hemoglobinopathies and hemoglobin biosynthesis.

A hemoglobinopathy can be defined as a hematological disorder caused by an alteration in the genetically determined molecular structure of a hemoglobin, resulting in a characteristic pattern of abnormalities in clinical laboratory findings and often, but not always, associated with overt anemia.[3] Strictly speaking, the thalassemias described in man cannot be related to an alteration in the primary structure of hemoglobin. However, they are considered to be hemoglobinopathies caused by alterations in the hemoglobin biosynthesis that result in quantitative differences in hemoglobin synthesis as part of the clinical picture.[4]

The human red cell normally contains a single major hemoglobin component A, and two minor components. The minor component called A_2 is associated with adult red blood cells. A 1–2 percent fetal hemoglobin component is due to the persistent production of gamma chains after the transition from fetal to adult hemoglobin. The occurrence of these minor components in all red blood cells of all individuals has been well documented.[4] The proportions of the components are usually maintained within certain definable limits that can be easily measured in the laboratory. Any alteration in the relative percentages of A_2 and/or F within an individual is often associated with a disease such as thalassemia.

In the study of control of the quantitative aspects of hemoglobin biosynthesis, animal models have been most useful.[1,2] Therefore, this paper will be devoted primarily to consideration of these models, and then will conclude by relating some general aspects of the primary structure of hemoglobin to its physiological function.

The quantitative distribution of the various major and minor hemoglobin components in adult human erythrocytes is fairly consistent, and variations can be used as a supportive finding for diagnosis of disease. On the other hand, certain variations in the proportions of hemoglobin components in other animal species seem to be a normal finding. The exact mechanism that controls these levels of quantitative distribution is of primary interest.

Certain phenomena related to quantitative control of hemoglobin synthesis have been studied in animals. One example is the dramatic shift from fetal hemoglobin to adult hemoglobin. Many studies using the hemoglobins of amphibians and chickens have been designed to increase the understanding of the mechanism causing the change from fetal to adult hemoglobin.[5,6,7] The mechanism behind this transition is of extreme importance because a thorough understanding of this phenomenon would provide the necessary knowledge to correct defects in individuals with abnormal hemoglobins by replacing the faulty and normal beta chains with the gamma polypeptide chain of the fetal hemoglobin type.

Several animal models appropriate for the study of the levels of hemoglobin synthesis can be identified (Table 1).

Several possible levels of control of hemoglobin synthesis that might be anticipated for all species have been listed in Table 2.

Figure 1 compares the occurrence of the polypeptide chains that form the hemoglobin types found in individual humans and animals during different stages of development.[8,9] The left-hand half of each graph represents the embryonic and fetal stages. Birth is indicated by a vertical line. Note that the gradual disappearance of human fetal hemoglobin due to cessation of the gamma chain production is accompanied by the gradual replacement with the beta chain, resulting in the gradual increase of the normal adult hemoglobin. The fetal hemoglobin level remains at one or two percent throughout nearly all

TABLE 1 Reference List of Animal Models[1,2]

Model	Experimental Use	Selected References
INVERTEBRATES		
1. *Daphnia*	Effects of vitamin B_{12} and other exogenous chemicals on hemoglobin synthesis	24
2. Nematoda	Effects of heme to stimulate hemoglobin synthesis	25
LOWER VERTEBRATES		
1. Fish		
a. Salmon	Effects of environment upon production of hemoglobin types (salt vs fresh water, aging, hormones	26
b. Ice Fish	Lack of hemoglobin	27
2. Amphibians		
a. Frogs	Transition of hemoglobin types during metamorphosis	7
b. Toads	Lack of hemoglobin	28
3. Avian		
a. Chicken	Effect of inorganic PO_4 on O_2 affinity, transition of embryonic $\rightarrow$ adult hemoglobin type	6, 41
HIGHER VERTEBRATES (MAMMALS)		
a. Sheep	Natural selection of hemoglobin types due to differences in physiological characteristics (oxygen affinity, Bohr effect)—Study of the switching mechanism for hemoglobin types A$\rightarrow$C	11, 12, 13, 14
b. Goat	Study of the switching mechanism of hemoglobin during development A$\rightarrow$C	9, 15
c. Cow	Heinz body formation due to toxins and oxidants	29, 30, 31
d. Horse	Genetic control of quantitative levels of hemoglobin types or components—Ambiguity of genetic code, multi-alleles	18, 19, 32

TABLE 1 Reference List of Animal Models (Continued)

Model	Experimental Use	Selected References
HIGHER VERTEBRATES (MAMMALS) continued		
e. Monkey	Genetic control of quantitative differences of major hemoglobin components within an individual— Models to study control of the synthesis of minor hemoglobin components such as A_2	8, 17, 33
f. Cat	Study of intraerythrocytic crystals	34
g. Mice and Rabbit	Ambiguity of genetic code vs multi-alleles	35, 36, 42, 43
h. Rat	Thalassemia	37
i. Deer	Quantitative control of hemoglobin components—Morphological erythrocyte forms (sickling of erythrocytes associated with various polymorphic hemoglobins)	38, 39, 40
j. Genet	Sickling erythrocytes	10, 40

TABLE 2 Control of Hemoglobin Biosynthesis

A. Transition of hemoglobins.
  Sequential transition of epsilon to gamma to beta polypeptide chains.
  Switching of fetal to adult hemoglobin as occurs in most mammalian species.
  Switching of adult A hemoglobin—Adult C hemoglobins in sheep and goats.
B. Heme = globin. Heme synthesis is coordinated with globin synthesis.
C. Equal number of alphas to non-alphas are synthesized per red blood cell.
D. Maximum hemoglobin concentration per cell is 34–36 percent MCHC.
E. In the case of the human heterozygote having a normal and an abnormal hemoglobin (in which the abnormal subunit is a product of the same alleles as the normal) these hemoglobins are *not* produced in a 1:1 proportion. By contrast, for most domestic animals in which polymorphic hemoglobins are the results of products of the same pair of alleles, the proportion of the hemoglobin components in the heterozygotes is 1:1.

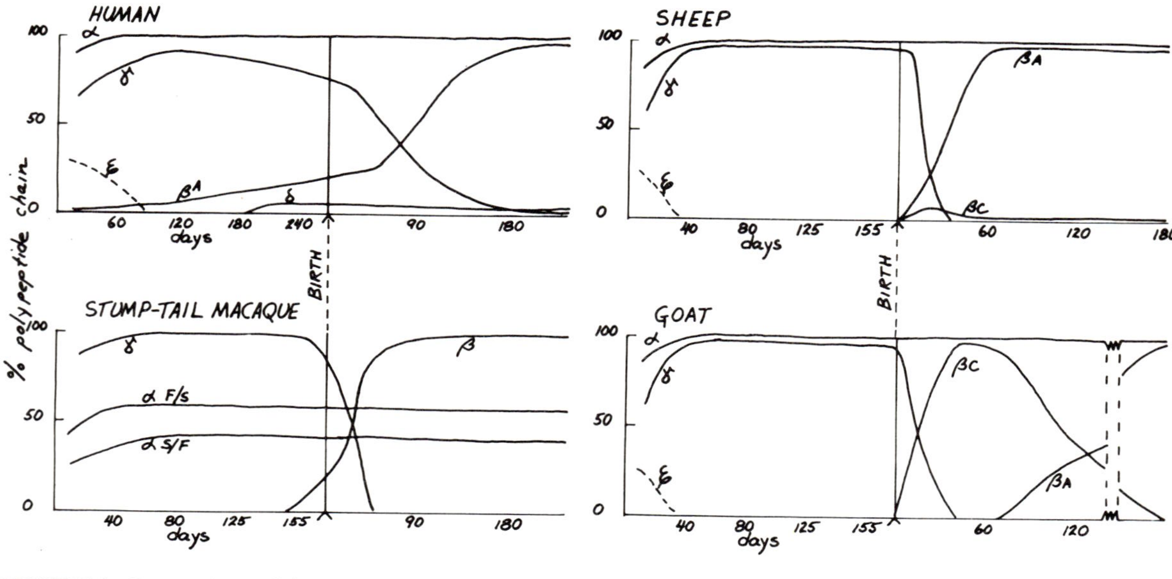

FIGURE 1   Proportions of the various hemoglobin polypeptide chains during early life for human, stump-tail macaque, sheep, and goat.[8, 9]

of the individual's life. In contrast to the human, the stump-tail macaque shows a rather rapid transition from fetal to adult hemoglobin with no fetal hemoglobin identifiable in the adult animals.[8]

A rapid transition also occurs in sheep and goats. This is in contrast to a rather prolonged period for the disappearance of fetal hemoglobin in the cow.[10] Another important phenomenon in sheep and goats is the appearance of hemoglobin C during development after birth. This is a hemoglobin that normally is not produced in healthy adult sheep and goats; it may be produced, however, under two conditions[11-14]: (1) during the normal sequence of hemoglobin production after birth, and (2) in adult animals having a severe anemia, whether from acute blood loss or chronic blood loss due to parasitism. Figure 2 demonstrates the change in an individual sheep of A phenotype in a period of 29 days to hemoglobin C after a serious blood loss. This phenomenon is reminiscent of the transition from fetal to adult hemoglobins, although it occurs at a different time in the animal's development.

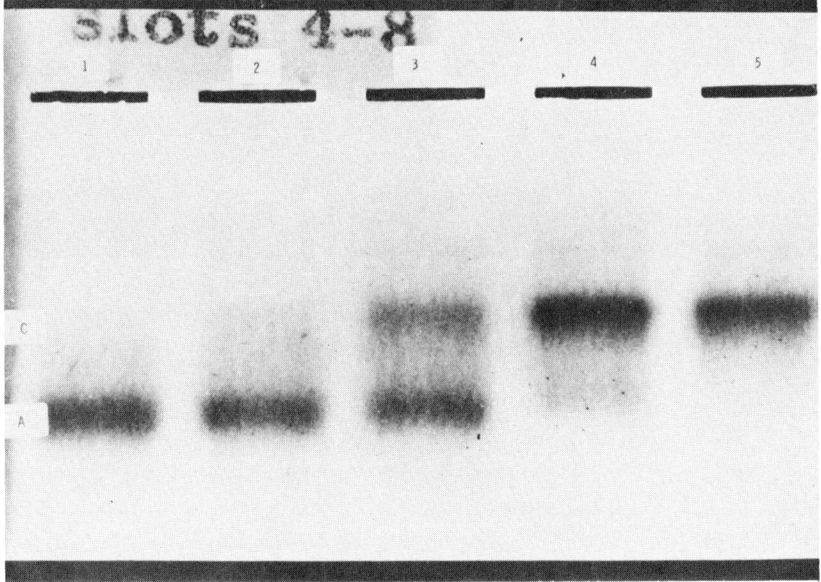

FIGURE 2   A demonstration on starch gel electrophoresis of the transition of sheep hemoglobin A to hemoglobin C in a single sheep phenotype A after acute hemolysis.[11] Shift in phenotype occurred in 29 days after initial doses of phenylhydrazine.

Therefore, it offers a useful model for study of the transition of one hemoglobin type to another. In fact, the transition of gamma to beta C to adult beta A in the sheep and goats may well represent the same type of transition that occurs from embryonic to fetal to adult hemoglobins. This is an ultimate level of differentiation that it would be important to understand.

In the first half of this presentation we have considered the switching of fetal to adult, or transition of the synthesis of various non-alpha polypeptide chains (Level A in Table 2). Now let us consider the quantitative control of various hemoglobin components as related to multiple hemoglobin, polymorphic hemoglobin or abnormal hemoglobin (Level E in Table 2).

"Multiple," "polymorphic," and "abnormal" are all terms that have been used to describe the existence of more than one hemoglobin with unique physical and chemical properties within a species. In this presentation the term *multiple* hemoglobins will be reserved for species whose members always have more than a single hemoglobin other than fetal. Such multiple components must each be genetically determined, and the separable components must have identical physical and chemical characteristics in every member of the species. The term *polymorphic* hemoglobins will be used in those cases where more than a single hemoglobin component exists within a species. Individual members within the species may have one or more of these physically and chemically distinct proteins. The frequency of occurrence for a polymorphic hemoglobin should be greater than would be expected from mutation alone. *Abnormal* hemoglobins will be an accepted term for variations in human hemoglobins, but within other species the term will be reserved for hemoglobins that exist in such low frequency as to suggest abnormality as a result of recent mutation, or that can be linked directly to a deleterious effect.

Figure 3 illustrates the proposal for the genetic control of polymorphic hemoglobins for cow, sheep, and goats. The vertical bars represent cistrons for the genetic control of the various polypeptide chain types that make the different hemoglobin components. In sheep of B phenotype the functional cistron for the beta C chain is absent—in contrast to goats, which regardless of their beta chain type have the cistron for the beta C chain.[9, 15]   In the chromosome of the cow, a homologous cistron for beta C is not evident. At the molecular level, the discovery of multiple amino acid substitutions, apparent upon comparing the structure of the beta chains of sheep and goat hemoglobins, the beta chains of sheep hemoglobins A and B, and the beta

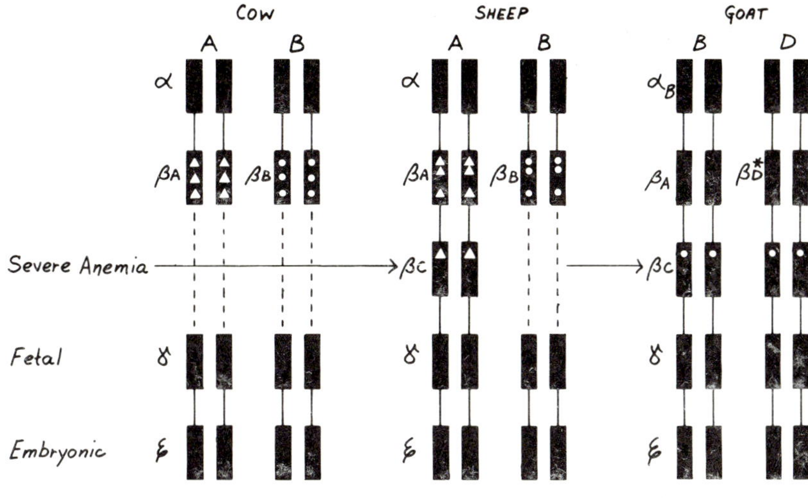

FIGURE 3   Proposal for the genetic control of the polymorphic hemoglobins of cow, sheep, and goat. The vertical bars represent cistrons controlling the production of the various polypeptide chains that combine as tetrameres to form the different hemoglobins. Three white circles or triangles in the bars indicate more than a single codon difference between cistrons of the same species. One white circle or triangle indicates a single condon difference between sheep beta C and goat beta C. Absence of white circles or triangles indicates identity of structure.

chains of goat hemoglobins A and D, is important. Yet, there is only a single amino acid difference between the structure of the beta C chain from sheep and the beta C chain from goats.

Figure 4 illustrates a proposal for the genetic control of the hemoglobin of sheep, man, and monkey. Here we have an example of polymorphic, abnormal, and multiple hemoglobins, respectively. In the human, the alpha chains combine with the various non-alpha chains to form the different hemoglobins illustrated in Figure 4—A_2, the normal adult A, the abnormal hemoglobin S, and the fetal form, or F. The heterozygous individual represented in the figure contains both the normal adult hemoglobin and an abnormal hemoglobin, Hb-S, which is associated with sickle cell anemia. The beta A and beta S chains differ by a single amino acid interchange. Note that the beta chains for the abnormal hemoglobins are not produced in a quantity equal to those for the normal hemoglobin.[16] The two hemoglobins (A and S) are in unequal quantities; this, therefore, represents a quantitative difference in production of the two beta chain types.

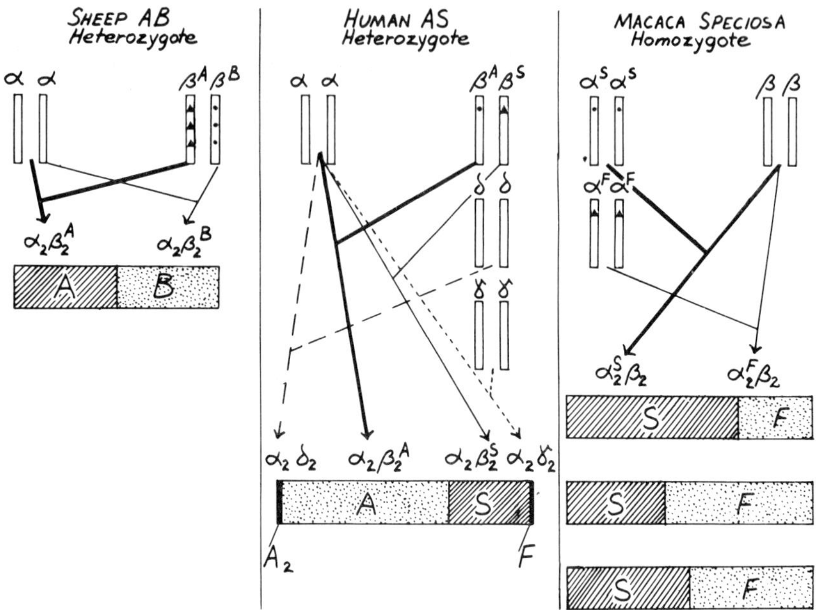

FIGURE 4   Proposal for the genetic control of the hemoglobins of sheep, human, and monkey. The vertical bars represent cistrons for the genetic control of the various polypeptide chains that make the different hemoglobins. The three solid triangles or solid dots emphasize multiple structural differences and are seen in the products of the single pair of cistrons for the sheep beta chains versus the single amino acid difference between the structure of the beta chains of human hemoglobins of A and S as represented by a single circle or triangle. Likewise, a single amino acid difference is seen between the two alpha chains that are products of two different pairs of cistrons in the stump-tail macaque.

The hemoglobins of heterozygous sheep AB, which have a common alpha chain, are produced in nearly equal concentrations. There are multiple structural differences between the A and B beta chains of the sheep, whereas a single structural difference is seen in the beta chain of most human hemoglobinopathies. Why do certain sheep, goats, and cows, which are heterozygous for hemoglobin type, produce these polymorphic hemoglobins in equal quantities, while the production of the two hemoglobins is unequal in other animals? In the stump-tail macaque two different alpha chain types and a structurally common beta chain exist in all individuals (Figure 4). In contrast to the sheep, in which the two hemoglobins are found in equal concentrations, the two hemoglobins of the stump-tail macaque may be in unequal concentrations, and three phenotypes may be represented based on the relative amounts of the two hemoglobins. Some animals have a distribution of 60/40 fast to slow; others exhibit 30/70 fast to slow, which is nearly the reverse[8, 17]; still others are 50/50. The bar graphs represents the relative amounts of the whole hemoglobins that are produced for the three phenotypes, as shown by the shaded or stippled areas (Figure 4).

This type of variation in the production of the multiple hemoglobin types in these homozygous monkeys is in contrast to that of some of the other animals. The question is, where do the control points occur at which the relative amounts of the hemoglobins are regulated? Table 3 gives a detailed outline of the various levels of control in protein synthesis. Table 3 is summarized graphically in Figure 5. An example of the influence of environment upon the activation of a new gene product is shown in the transition from sheep hemoglobin A to hemoglobin C.

Three important examples of genetically determined differences that are responsible for the relative production of the hemoglobin

TABLE 3   Molecular Levels of Quantitative Control of Hemoglobin Synthesis

---

1. Replication
   A. Duplication of a gene
   B. Mutation in a gene
2. Transcription
   A. Unequal number of alleles
   B. Unequal transcription of two alleles
   C. Ambiguous codon, recognizes two t-RNA's
   D. Limited supply of rare t-RNA
   E. Ambiguous enzyme—one t-RNA for two amino acids
   F. Environment
   G. Different alleles yielding electrophoretically identical products
3. Translation
   A. Unequal number of m-RNA
   B. Unequal stabilities of m-RNA
   C. Unequal rates of assembly—initiation, translation, release
4. Association
   A. Unequal rates of combination between multiple alpha or beta chains with a single beta, gamma or alpha chain.
5. Destruction
   A. Unstable hemoglobin, removal of cells with high concentrations.

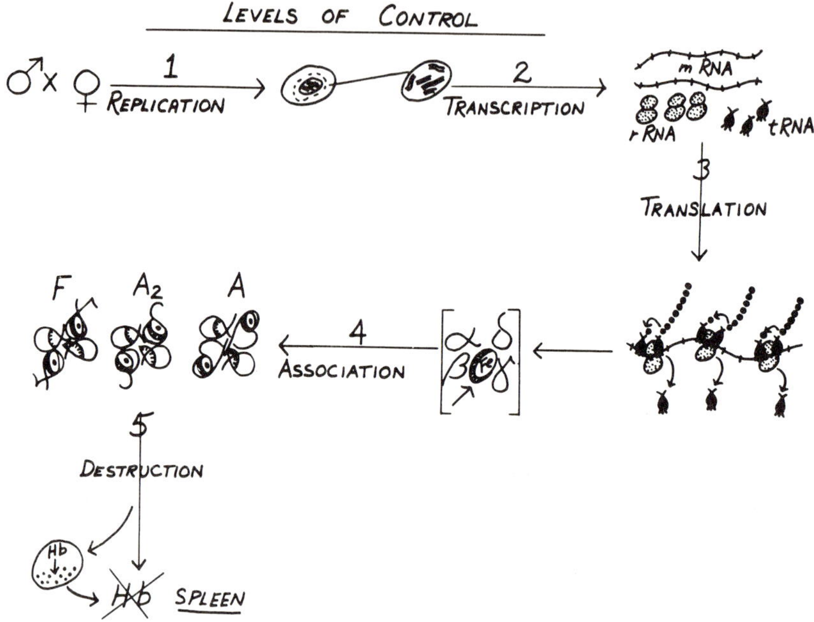

FIGURE 5   Levels of quantitative control for hemoglobin biosynthesis. The numbers refer to the levels of control listed in Table 3.

components are given in Figure 6. The stump-tail macaque, horse, and goat hemoglobins are illustrated by their starch gel behavior in this figure.[17-19] The electrophoretic behavior of horse and monkey hemoglobins have been illustrated with the various polymorphic hemoglobins of goats, since the polymorphic hemoglobins of goats also show a variety of quantitative differences.[1] Note the stump-tail macaque individual with a fast component of only 30 percent and a slow component of 70 percent, and then the individual with a nearly reverse ratio—60 percent fast and 40 percent slow. These differences are determined genetically rather than by environmental influences. Among horses, there are individual animals for which the slow component is 30 percent and the major component is 70 percent, and others for which the minor, or slow, component is 10 percent.

For simplicity, in the first reference, the authors have not considered the well-established evidence for duplication of the alpha-chain cistrons as a possible explanation for the quantitative differences in the polymorphic hemoglobin of goats. It would be well to

consider that, where there is evidence for cistron duplication, the polypeptide chain types are produced in unequal proportions, but that
where the genetic evidence indicates a single cistron pair, the hemoglobin types are nearly equal in proportion. Both of the above explanations are necessary to explain the genetic control in the goat.
The complexity of this model for goats precludes illustration here.
For an excellent discussion of genetic control in the goat, the reader
is referred to the papers of Huisman *et al.*[9, 15]

In almost every case where there is a reversal of major and minor
components, or where quantitative differences exist in the major and
minor components, these examples could be explained by duplication
of genetic material. That is, more than a single pair of alleles are pro-

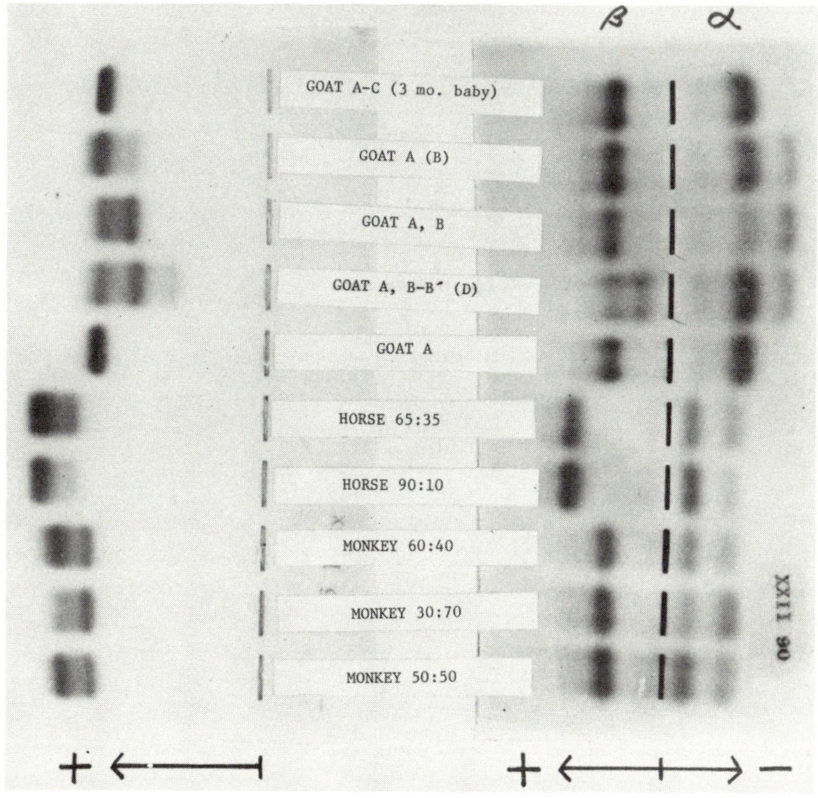

FIGURE 6   Left-hand side: Starch gel electrophoresis at pH 8.6. Right-hand
side: Separation of the alpha and beta chains in starch urea gel at pH 8.1.

ducing similar alpha-like or beta-like chains. Both cases differ from that of sheep having hemoglobin A and B, previously presented, because the polymorphic hemoglobin A and hemoglobin B contain different beta chains that are products of the same pair of alleles. In the multiple hemoglobins of the monkey, however, the alpha chains are due to duplication of the alpha cistron, so that there must be at least two pairs of alpha alleles. Figure 7 is based upon evidence recently given by Schroeder *et al.*,[20] that there are multiple gamma alleles for the gamma chains of the human fetal hemoglobins and not just a single pair of alleles, as previously accepted.

Certainly, when multiple alleles are present, there is an additional level of control in transcription (Figure 5) in which more messengers of one globin type than of another can be produced, because of the unequal numbers of pairs of alleles for the two hemoglobins. Because the number of allelic pairs for the two types is unequal, there would

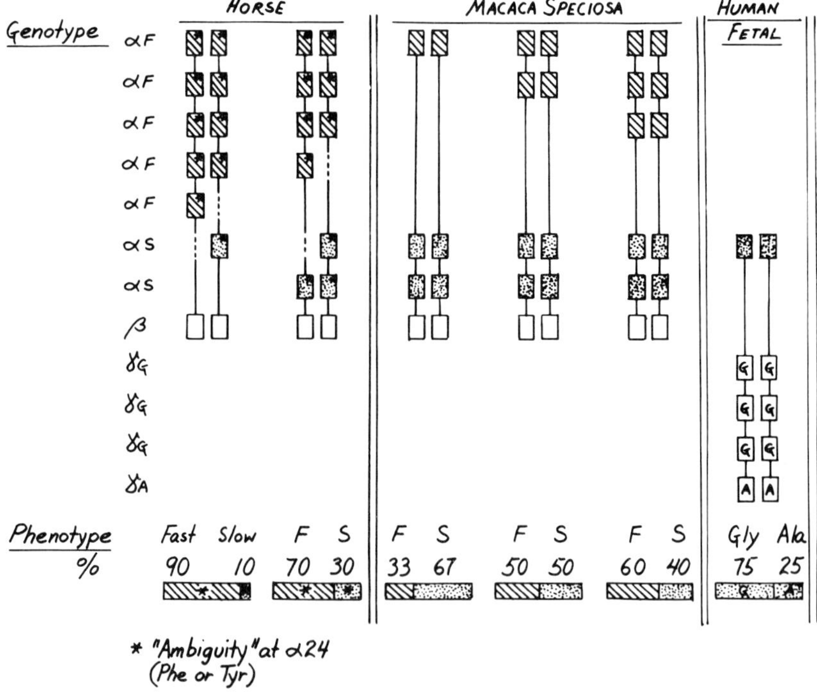

FIGURE 7 A theoretical proposal for an explanation of quantitative differences in multiple hemoglobin components of the horse and the stump-tail macaque, as based on a similar proposal by Schroeder for human fetal hemoglobins.[20]

be indeed a quantitative difference in these two hemoglobin types. At this level of control there are unequal numbers of allelic pairs for the two variations of the same fetal hemoglobin. This finding can be extended to give a proposed level of control that would explain some of the quantitative differences demonstrated in the various monkey and horse hemoglobins (Figure 7).

The work of Schroeder et al.[20] has shown that there may be as many as four pairs of alleles for the gamma chain in human fetal hemoglobin. Their evidence is based upon the finding of a quantitative difference in the two phenotypes of fetal hemoglobin caused by a single amino acid difference (glycine for alanine). Note the quantitative difference in the fetal hemoglobin that has glycine at a given sequence position versus the fetal hemoglobin having alanine at the same sequence position. This quantitative difference is explained by the number of alleles or pairs of cistrons involved. Now, by applying this principle of multiple alleles, a proposed model can be given for the explanation of the quantitative differences demonstrated for the horse and the stump-tail macaque.

In conclusion, four major points should be emphasized:

1. The use of animal models for study of the transition of hemoglobin types was illustrated (Figures 1, 2). The initiating mechanism that is responsible for the change in the synthesis from one polypeptide chain type to another is very complex. Does this change in hemoglobin types occur with a change in cell populations and cell types, and with a concomitant shift in enzymes? Is this change hormone mediated? A knowledge of this mechanism certainly has implications for advanced therapy.

2. The control levels that result in quantitative differences in hemoglobin components, particularly in the heterozygote, have been considered (Table 2E, Figure 4). It is an interesting fact that in most of the human hemoglobin abnormalities, in which the structural difference is due to a single amino acid substitution, there has been a subsequent reduction of the relative amount of this abnormal hemoglobin in the heterozygous individual. This finding of quantitative differences in the synthesis of the abnormal versus the normal has led Itano to formulate the structure rate hypothesis that a single amino acid substitution can alter the net rate of production of the polypeptide chain in which it occurs.[21] For animals having polymorphic hemoglobins, this heterogeneity is due to genetic control by a single pair of alleles. In the beta chains of sheep, for example, these structural differences

are due to multiple amino acid substitutions; yet these globins are produced in a 1:1 proportion. This might implicate still another level of selection for functional proteins.

Most of the abnormal hemoglobins of man, except for the S hemoglobin associated with sickle cell anemia, occur infrequently. This suggests that these structurally different proteins do not represent evolutionarily effective amino acid substitutions. In the case of animals that are heterozygous for hemoglobins, the frequencies of their hemoglobins in the population suggest that these amino acid substitutions have been evolutionarily effective. It is assumed that a majority of evolutionarily effective substitutions have to be considered on two different criteria: first, that the amino acid substitutions should not critically alter the functional properties of the protein and, second, that the substitutions do not alter the regulation of the synthetic activity, since the regulation of synthetic activity may be dependent on the structural gene.

3. The third major point concerns animals in which hemoglobin heterogeneity can be explained by multiple alleles (Figure 7). The duplication of alpha chains such as occurs in the goat, the monkey, and the horse is an example. In cases in which multiple alleles dictate the heterogeneity of a hemoglobin, and in which all multiple allelic forms do occur in all the individuals of the species, structural substitutions cannot entirely explain the quantitative differences in hemoglobin proteins. Indeed, a variation in the number of alleles may explain these quantitative differences.

4. The fourth item of particular importance is the finding that the evolutionarily effective amino acid substitutions occurring in the sheep have produced no alteration of the synthetic ratios of the two structurally different beta polypeptide chains (Figure 4). However, there certainly has been a change in physiological properties; for example, the sheep hemoglobin A has an O_2 dissociation curve to the left of that of the B sheep hemoglobin. This results in an increase in oxygen affinity for adult sheep hemoglobin A. Thus, the sheep with hemoglobin A have been referred to as high altitude animals. It is interesting to consider the results of Evans and Whitlock,[22] which indicate that the hemoglobin concentration and hematocrits for sheep having hemoglobin A are significantly higher than for the sheep having hemoglobin B in the New York State area. This suggests that perhaps the A sheep, which are referred to as "high altitude animals," at low altitude are anoxic at the tissue level as compared to the "low land" B sheep. An experiment would be desirable in which the B sheep and the A sheep

could be studied at both high and low altitudes to see if there is a reversible shift in average hematocrit.

Recent studies of the human hemoglobinopathies referred to as hemoglobin Rainer and Yakima[23] lend support to this possible explanation of a relation of hemoglobin function to the number of erythrocytes and thus the hematocrit and hemoglobin concentrations. These hemoglobin abnormalities, due to a single amino acid interchange, have abnormal oxygen dissociation curves, with an increase in oxygen affinity. The affected individuals have an erythrocytosis due to this altered physiological function of the hemoglobin molecule. In conclusion, it may be stated that the understanding of disease in man often leads to the understanding of phenomena in animals.

ACKNOWLEDGMENTS

This investigation and presentation was supported by the National Science Foundation Research Grant GB-7876, United States Public Health Service Grants H-5004, 5-PO6-FR00366-03 and Florida Heart Association Grant 68 AG 8.

Dr. H. Kitchen is the recipient of a Research Career Development Award (K3-AM-31, 811) from the National Institute of Arthritis and Metabolic Diseases and is presently an Associate Professor, Michigan State University. A portion of this investigation was performed by Ethel Oliver Kouba, as part of Ph.D. dissertation, "Study of Multiple Hemoglobins in *Macaca speciosa*," University of Florida, August, 1969.

The authors wish to thank Sarah L. Martin, Ruth Martin, and John Neel, for technical assistance and Kathleen Genser and Yvonne Kitchen for assistance in preparing and typing this manuscript.

REFERENCES

1. Kitchen, H. 1968. Comparative biology: Animal models of human hematologic disease. Pediat. Res. 2(3):215.
2. Kitchen, H. 1969. Heterogeneity of animal hemoglobins. Advances in Veterinary Science, C. A. Brandly and C. E. Cornelius editors, Vol. 13, Academic Press, New York.
3. Dorland's Illustrated Medical Dictionary, 24th edition, W. B. Saunders Co., Philadelphia, Pennsylvania.
4. Wintrode, W. M. 1967. Clinical hematology, 6th edition, Lea and Febiger, Philadelphia, Pennsylvania.

5. Fraser, R. C. 1966. Polypeptide chains of chick embryo hemoglobin. Biochem. Biophys. Res. Commun. 25:142–146.

6. Wilt, F. H. 1968. The control of embryonic hemoglobin synthesis. *In* Advances in Morphogenesis 6:89–128, Abercrombie, M. and Brachet, J., editors, Academic Press, New York.

7. Moss, B., and V. M. Ingram. 1968b. Hemoglobin synthesis during amphibian metamorphosis II. Synthesis of adult hemoglobin following thyroxine administration. J. Molec. Biol. 32:493–504.

8. Kitchen, H., J. W. Eaton, and V. G. Stenger. 1968. Hemoglobin types of adult, fetal and newborn subhuman primates: *Macaca speciosa.* Arch. Biochem. Biophys. 123:227.

9. Huisman, T. H. J., J. P. Lewis, M. H. Blunt, H. R. Adams, A. Miller, Andree M. Dozy, and Evalyn M. Boyd. Hemoglobin C in newborn sheep and goats: A possible explanation for its function and biosynthesis. Pediat. Res. In press.

10. Kitchen, H. Observation in this laboratory.

11. Kitchen, H., J. W. Eaton, and W. J. Taylor. 1968. Rapid production of a hemoglobin by induced hemolysis in sheep: Hemoglobin C. Amer. J. Vet. Res. 29:2, 281.

12. Boyer, S. H., P. Hathaway, F. Pascasio, C. Orton, J. Bordley, and M. A. Naughton. 1966. Hemoglobins in sheep: Multiple difference in amino acid sequences of three beta chains and possible origins. Science 153:1539.

13. van Vliet, G., and T. H. J. Huisman. 1964. Changes in the hemoglobin types of sheep as a response to anaemia. Biochem. J. 93:401.

14. Moore, S. L., W. C. Godley, G. van Vliet, J. P. Lewis, E. Boyd, and T. H. J. Huisman. 1966. The production of hemoglobin C in sheep carrying the gene for hemoglobin A: Hematologic aspects. Blood 28:314.

15. Adams, H. R., R. N. Wrightstone, A. Miller, and T. J. J. Huisman. 1969. Quantitation of hemoglobin a-chains in adult and fetal goats; gene duplication and the production of polypeptide chains. Arch. Biochem. and Biophys. 132:223–236.

16. Boyer, S. H., P. Hathaway, and M. D. Garrick. 1964. Modulation of protein synthesis in man: An *in vitro* study of hemoglobin synthesis by heterozygotes. Cold Spring Harbor Symp. Quant. Biol. 39:333.

17. Oliver, E., and H. Kitchen. 1968. Hemoglobins of adult *Macaca speciosa*: An amino acid interchange [a-15 (gly-asp)]. Biochem. Biophys. Res. Commun. 31:5, 749.

18. Kitchen, H., W. F. Jackson, and W. J. Taylor. 1966. Hemoglobin and hemodynamics in the horse during physical training. Proceedings of the American Association of Equine Practitioners, 11th Session, Miami, Florida, 1965:97–110.

19. Kitchen, H., and C. W. Easley. Comparison of multiple hemoglobins from genus *Equus.* Fed. Proc. 27.

20. Schroeder, W. A., T. H. J. Huismann, J. Roger Shelton, J. B. Shelton, Enno F. Kleihauer, A. M. Dozy, and Barbara Robberson. 1968. Evidence for multiple structural genes for the γ chain of human fetal hemoglobin. Proc. Nat. Acad. Sci. U.S. 60:537.

21. Itano, H. A. 1965. *In* Abnormal hemoglobins in Africa, J. H. P. Jonxis, editors, Blackwell Scientific Publications, Oxford.

22. Evans, J. V., and J. H. Whitlock. Genetic relationship between maximum hematocrit values and hemoglobin type in sheep. Science 145.
23. Jones, Richard T. 1968. Mechanisms underlying polymorphisms in hemoglobins and other proteins. *In* Plenary Session Papers XII, Congress International Society of Hematology 64–716.
24. Fox, H. M., and E. A. Phear. 1953. Factors influencing haemoglobin synthesis by *Daphnia*. Proc. Roy. Soc. London, Ser. B, 141:179–189.
25. Lee, D. L., and M. H. Smith. 1965. Hemoglobins of parasitic animals. Exp. Parasitol. 16:392–424.
26. Vanstone, W. E., E. Roberts, and H. Tsuyuki. 1964. Changes in the multiple hemoglobin patterns of some Pacific salmon genus oncorhynchus during the parr-smolt transformation. Can. J. Physiol. Pharmacol. 42:697.
27. Ruud, J. T. 1954. Vertebrates without erythrocytes and haemoglobin. Nature 173:848–850.
28. Ewer, D. W. 1959. A toad (Xenopus laevis) without hemoglobin. Nature 183:271.
29. Clegg, F. G., and R. K. Evans. 1962. Hemo-lobinemia of cattle associated with the feeding of Brassical species. Vet. Res. 74:1169.
30. Dunbar, G. M., and T. A. M. Chambers. 1963. Suspected kale poisoning in dairy cows. Vet. Res. 75:566.
31. Greenhalgh, J. F. D., G. A. M. Sharman, and J. N. Aitken. 1969. Kale Anemia. 1. The toxicity to various species of animals of three types of kale. Res. Vet. Sci. 10:1.
32. Kilmartin, J. V., and J. B. Clegg. 1967. Amino acid replacement in horse haemoglobin. Nature 213:269–271.
33. Boyer, S. H., E. F. Crosby, G. L. Fuller, A. N. Noyes, and J. G. Adams. The structure and biosynthesis of hemoglobin A and A_2 in New World primate, *Ateles paniscus*, I. A. Preliminary Account. Ann. N.Y. Acad. Sci. In press.
34. Altman, Norman H. Intraerythrocytic hemoglobin crystals in cats and their comparison with hemoglobinopathies of man. Presented at the American Society of Veterinary Clinical Pathologists, p. 34. The Scientific Proceedings of the 106th Annual Meeting, American Veterinary Medical Association, July 13–17, 1969. Minneapolis, Minnesota.
35. Rifkin, D. B., D. I. Hirsh, M. R. Rifkin, and W. Konigsberg. 1966. A possible ambiguity in the coding of mouse hemoglobin. Cold Spring Harbor Symp. Quant. Biol. 31:715–718.
36. Von Ehrestein, G. 1966. Translational variations in the amino acid sequence of the α-chain of rabbit hemoglobin. Cold Spring Harbor Symp. Quant. Biol. 31:705–714.
37. Sladic-Simic, D., P. N. Martinovitch, Zivkovic, D. Pavic, J. Martinovic, M. Kahn, and H. M. Ranney. 1969. A thalassemia-like disorder in Belgrade laboratory rats. Ann. N.Y. Acad. Sci. In press.
38. Kitchen, H., F. W. Putnam, and W. J. Taylor. 1967. Hemoglobin polymorphism in white-tailed deer: Subunit basis. Blood 29:867.
39. Kitchen, H., C. W. Easley, F. W. Putnam, and W. J. Taylor. 1968. Structural comparison of polymorphic hemoglobins of deer with those of sheep and other species. J. Biol. Chem. 243(6):1204–1211.

40. Gulliver, G. 1840. Observations on certain peculiarities of form in the blood corpuscles of the mammiferous animals. Proc. Zool. Soc. London 17:325–327.
41. Huisman, T. H. J., and J. M. S. Van Veen. 1964. Studies on animal hemoglobins. III. The possible role of intercellular inorganic phosphate on the oxygen equilibrium of the hemoglobin in the developing chicken. Biochim. Biophys. Acta 88:367–374.
42. Hilse, Kurt, and R. A. Poss. 1968. Gene duplication as the basis for amino acid ambiguity in the alpha chain polypeptides of mouse hemoglobins. Proc. Nat. Acad. Sci. U.S. 61:930–936.
43. Ishibashi, S., L. C. Niu, Y. H. Yuh, M. C. Niu, and E. B. Kalan. 1968. Evidence for dimorphism in rabbit hemoglobins. Proc. Soc. Exp. Biol. Med. 128:3, 879–885.

Cyclic Neutropenia in Man and Dog

JOHN E. LUND

Cyclic neutropenia (C-N) was first observed by Leale,[1] in 1910, in a 3½-month-old infant. The child presented with recurrent episodes of furnuculosis, fever, oral ulcers, and "an unusual blood picture." The unusual blood picture was a recurrent neutropenia that repeated at 21-day intervals. The disease persisted until the patient died at age 25.[2-4] In 1946, Valquist and Verneholt[5] attempted to classify this rare syndrome as a disease entity. They collected five case reports from the literature and added one of their own. Reimann and de Berardinis[4] assembled 16 cases—13 from the literature, 2 from personal communications (subsequently published as case reports), and 1 of their own. They suggested that the disease was a specific entity and should be grouped with other periodic disorders, all characterized by the apparent absence of provocation and the appearance of specific symptoms at regular intervals. Reimann later[6] published a monograph on periodic diseases in which he described 42 persons with cyclic neutropenia. Reimann used the term "periodic myelodysplasia" to describe the disease, and, although in many ways it is a more meaningful term than cyclic neutropenia, it has not gained acceptance.

HUMAN CYCLIC NEUTROPENIA

In a recent review,[7] 59 complete descriptions of human C-N patients were found. Eleven other patients were classified as possible C-N cases. (Most were categorized as questionable cases because neither the criteria for the diagnoses nor the data from serial leukocyte counts were

71

presented in the reports.) If all of the latter individuals had C-N, there have been approximately 70 human cases reported since 1910. Thirty-three of the C-N patients were women.

The symptoms of the disease were detected in 32 of the patients when they were less than three years old. Sixteen patients were between one and 10 years of age when the symptoms of the disease became apparent, five between the ages of 11 and 20 years, seven between the ages 21 and 40 years, and six were over 40 years old. The age of onset for one patient[8] was not determined.

In 51 of the patients, the approximate length of the disease cycle was 21 days; six had cycles of approximately 14 days; six had cycles near 28 days in length; and Laske *et al.*[9] reported a cycle length of 40 days in a 7-year-old boy. The longest reported cycle occurred in a 42-year-old woman whose symptoms began at age 34.[10, 11] Initially the episodes were widely separated but the interval gradually shortened to 6 to 8 weeks. The age of the patients at the onset of the disease had no apparent relationship to the length of the observed neutrophil cycle.

Other than the changes in the blood and bone marrow, no specific lesions have been found in biopsy specimens taken from C-N patients or in specimens obtained from postmortem examination. Two of the patients had lymphosarcoma along with cyclic neutropenia.[12, 13] Four patients died of pneumonia[4, 14–16]; one[17] died from gross hematemesis resulting from gastric ulceration; one had cecitis, ileitis, and local peritonitis[18]; and one[19] was not examined postmortem.

CANINE CYCLIC NEUTROPENIA

In 1967, we[20] described the occurrence of C-N in three gray Collie dogs. The neutropenic episodes occurred at 11-day intervals and lasted from two to four days. The course of the disease in the dogs was characterized by severe infections during the neutropenic episodes, high temperature, arthralgia, and death at an early age. Except for the apparent difference in neonatal mortality and the difference in the length of the disease cycle, the course of the disease was quite similar to that observed in man. The gray color in the Collie with C-N is a specific color dilution in the Collie breed and should not be confused with other color variants.[21] The C-N disease and gray coat color are inherited in a simple autosomal recessive pattern, and are the result either of a single genetic defect or of two closely linked traits.[7]

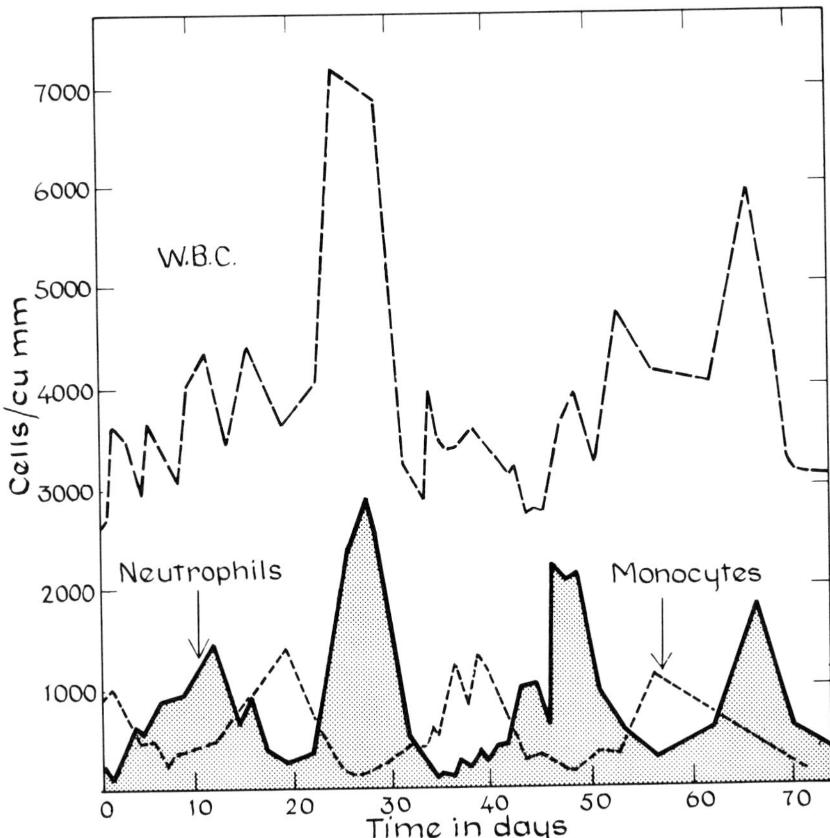

FIGURE 1   Variation in the number of peripheral blood neutrophils from a 12-year-old girl (from Lorré and Denys[30]). Note the low neutrophil count between episodes of neutropenia, a common finding in many of the human patients.

We[22] recently reported that erythrocytes also are not produced in the bone marrow during the period of neutrophil production failure. Examination of bone-marrow smears during the interval when the failure of neutrophil production was evident (e.g., just prior to peripheral neutropenia) indicated that both neutrophil and erythrocyte production was depressed. Because of technical difficulties in obtaining accurate bone-marrow cell counts, erythrocyte production was evaluated by comparing the relative birth rates of the neutrophils and erythrocytes. The relative birth rate did not shift significantly in favor of

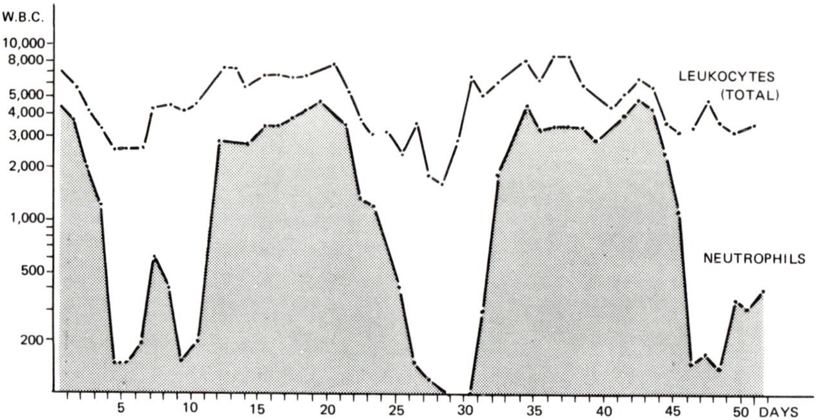

FIGURE 2   Case 1 reported by Videbaek[26] illustrating the pronounced cyclic variation in the neutrophil granulocyte and leukocyte counts during a period of 51 days. The neutrophil count between neutropenic episodes is nearly normal.

erythropoiesis during the initial neutrophil production failure, which indicated that erythrocyte production was also depressed.

The dogs with C-N develop a microcytic, hypochromic anemia and have retarded growth and sexual development.[7] These deficiencies may be attributed principally to the effects of repeated infections during the neutropenic periods that, because of the short interval between episodes, simulate chronic infection.

Most C-N dogs die before six months of age as the result of overwhelming infection, particularly pneumonia and gastroenteritis.[7, 23] Amyloid deposition in the visceral organs is noted in C-N dogs from 4 months to 2½ years of age.[7, 23] The severity of the amyloid deposition depends both on the age of the dog and on the severity of infection incurred during the neutropenic episodes.

NATURE OF THE CANINE C-N DEFECT

The cyclic variation of neutrophils in C-N dogs has been determined to result from the periodic failure of bone marrow to produce cells. The cyclic variation could result from a periodic change in the rate of utilization of neutrophils or from changes in the rate of release of mature cells from the bone marrow granulocyte reserve (MGR). These possibilities have been eliminated, however, by the demonstration that

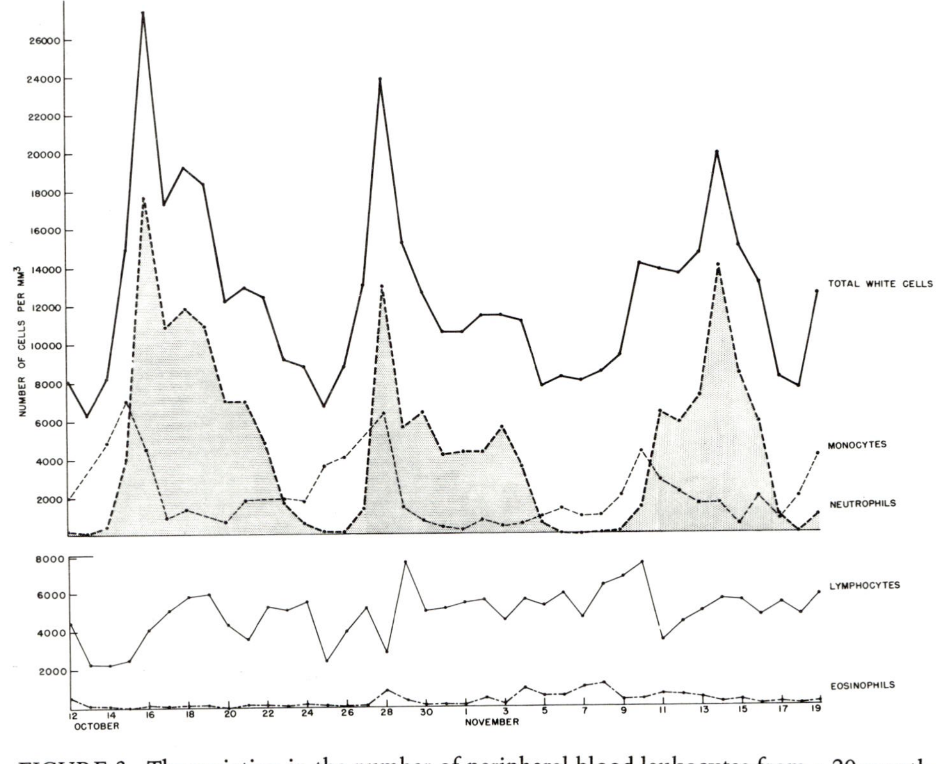

FIGURE 3 The variation in the number of peripheral blood leukocytes from a 20-month-old male C-N dog. Compensatory monocytosis is evident during the 3 neutropenic episodes.

the half-life of neutrophils in the blood of C-N dogs when the cell count is falling is similar to that of normal dogs, and that the MGR is depleted prior to peripheral neutropenia.[7] The sequential loss and recovery of bone-marrow cell development involving both myeloid and erythroid elements suggest that the disease may arise from a periodic failure of stem-cell differentiation.[7]

COMPARISON OF CANINE AND HUMAN C-N

The clinical manifestations in the two species are very similar. Both man and dog have frequent (often severe) bacterial infections, fever, malaise, and arthralgia. Other ancillary manifestations in C-N dogs and several human patients include anemia and thinning of the cortices of the long bones.[7]

Retardation of growth and sexual development has not been reported in human patients but is consistently observed in the dogs.[7, 23] The short disease cycle in dogs often results in clinical illness 3 to 5 days out of every 10, and in some instances clinical recovery is evident for only 1 or 2 days or not at all. Since skeletal dwarfism and infantilism have been associated with chronic diseases in man, it may be that the frequent infections observed in C-N dogs simulate chronic illness. However, the possibility of a direct effect of the genetic defect on growth and development has not been eliminated.

Early death characterizes the canine disease, whereas many human patients survive to adulthood. This may be because the short cycle in dogs does not allow sufficient time for complete recovery from infection, and may, therefore, tend to promote development of fatal disease, and account for the discrepancy in death rates between the two species. The difference, however, may not be real, or may be less than appears at a casual glance. A high death loss, which would not be readily recognized, could occur among infant C-N patients since, unlike the C-N dogs, they lack a coat color "marker."

Canine C-N is inherited as a simple autosomal recessive trait.[7] More than one-half of the human patients reported exhibited overt manifestations of disease during infancy, indicating a strong probability that the disease may be congenitally acquired. That the disease is heritable is indicated by five reports of familial C-N.[16, 24-27]

Many of the human C-N patients were reported to have a "maturation arrest" in the neutrophil developmental sequence; it is probable, however, that this impression is in error and results from an infrequent bone marrow sampling regimen.[7] The investigation by Page and Good[28] is an exception, but they postulated two defects, a failure of stem cell differentiation and a maturation defect at the promyelocyte stage. Other investigators who examined their patients by the use of sequential bone-marrow samples found no indication of maturation arrest. These latter findings agree with those derived from studies of C-N dogs, although an erythrocyte production defect has not been demonstrated in the human patients. Morley *et al.*[27] observed fluctuations of serum iron in two of their patients and concluded that erythrocyte production was cyclic; however, the possibility of iron deficiency secondary to infection[29] was not considered and bone marrow studies were not reported.

There is a possibility that erythrocyte production in man is affected, but it is not readily apparent in the peripheral blood because of the slow turnover of red blood cells. Conclusive evidence may not be attainable at present because of the inadequacy of routine techniques for determining total bone-marrow cell counts. The special techniques required to study the disease are not readily accepted by human patients, and very few patients are available.

CONCLUSION

Human cyclic diseases have been observed and studied for many years but their pathogenesis is not known and specific treatment is not available. For example, the understanding of C-N has not advanced significantly since Leale[1] described the first case in 1910. With the discovery of canine C-N disease, the first recorded cyclic disease in animals, it is now possible to make a definitive study with an animal model. It may also be possible to gain a better understanding of cyclic diseases in general and obtain insight into more useful methods for study and control of comparable human diseases.

REFERENCES

1. Leale, M. 1910. Recurrent furunculosis in an infant showing an unusual blood picture. JAMA. 54:1854.
2. Rutledge, B. H., O. C. Hansen-Prüss, and W. S. Thayer. 1930. Recurrent agranulocytosis. Bull. Johns Hopkins Hosp. 46:369.

3. Thompson, W. P. 1934. Observations on the possible relationship between agranulocytosis and menstruation with further studies on a case of cyclic neutropenia. New Eng. J. Med. 210:176.
4. Reimann, H. A., and C. T. de Beradinis. 1949. Periodic (cyclic) neutropenia, an entity. A collection of sixteen cases. Blood. 4:1109.
5. Vahlquist, B. 1946. Agranulocytose im kindesalter. Ziet. Kinder. 96:106.
6. Reimann, H. A. 1963. Periodic diseases. F. A. Davis Co., Philadelphia.
7. Lund, J. E. 1969. Canine cyclic neutropenia. Doctor of Philosophy Thesis, Washington State University, Washington.
8. Cohen, D. W., and A. L. Morris. 1961. Periodontal manifestations of cyclic neutropenia. J. Periodontol. 32:159.
9. Laski, B., A. Sass-Kortsak, and P. A. Hillman. 1954. Cyclic neutropenia and agammaglobulinemia. Am. J. Dis. Child. 88:820.
10. Dörken, H. 1952. Beitrag zum krankheitsbild der zyklischen agranulocytose. Schweiz. Med. Wschr. 43:1123.
11. Dörken, H. 1954. Lymphosarkom-Leukämie, cyclische agranulocytose und latente neutropenie in einer familie. Klin. Wochen. 32:173.
12. Sandella, J. F. 1951. Cyclic acute agranulocytosis: Report of a case with improvement after splenectomy. Ann. Int. Med. 35:1365.
13. Natelson, R. P. 1953. Cyclic neutropenia with giant follicular lymphoblastoma and lymphosarcoma. Report of a case with splenectomy. Blood. 8:923.
14. Good, R. A., and R. L. Varco. 1955. A clinical and experimental study of agammaglobulinemia. J. Lancet. 75:245.
15. Owren, P. A. 1949. Cyclic agranulocytosis. Acta Med. Scand. 134:87.
16. Becker, F. T., W. D. Coventry, and J. L. Tuura. 1959. Recurrent oral and cutaneous infections associated with cyclic neutropenia. Arch. Dermatol. 80:731.
17. Rolland, C. F., and L. S. P. Davidson. 1950. A case of cyclic neutropenia. Edinburgh. Med. J. 57:205.
18. Barsby, B. E., and H. G. Close. 1942. Recurrent neutrophil agranulocytosis. Lancet. 1:99.
19. Hlavoň, J. 1964. Vzáchá forma vrozené cellulárni nedostatecnosticylkicá neutropenie. Cesk. Pediat. (Praha). 19:152.
20. Lund, J. E., G. A. Padgett, and R. L. Ott. 1967. Cyclic neutropenia in grey Collie dogs. Blood 29:452.
21. Lund, J. E., J. R. Gorham, and G. A. Padgett. 1969. Canine cyclic neutropenia: Diagnosis and treatment. Ved. Med. Rev. (in press).
22. Lund, J. E., and M. A. Maloney. 1969. Erythrocyte production in canine cyclic neutropenia. Fed. Proc. 28:685.
23. Cheville, N. F. 1968. The gray Collie syndrome. J. Amer. Vet. Med. Ass. 15:620.
24. Hahneman, B. M., and H. L. Alt. 1958. Cyclic neutropenia in a father and daughter. JAMA. 168:270.
25. Gorlin, R. J., and A. P. Chaudhry. 1960. The oral manifestations of cyclic (periodic) neutropenia. Arch. Dermatol. 82:344.
26. Videbaek, A. 1962. Cyclic neutropenia. Report on three cases. Acta Med. Scand. 172:715.

27. Morley, A. A., J. P. Carew, and A. G. Baikie. 1967. Familial cyclical neutro-
    penia. Brit. J. Haematol. 13:719.
28. Page, A. R., and R. A. Good. 1957. Studies on cyclic neutropenia. A.M.A.J.
    Dis. Child. 94:623.
29. Cartwright, G. E. 1966. The anemia of chronic disorders. Seminars Hematol.
    3:351.
30. Lorré, J., and P. Denys. 1960. Contribution à l'étude des neutropénies
    cycliques. Acta Pediat. Belg. 14:178.

The Ehlers-Danlos Syndrome
of Dogs and Mink

GERALD A. HEGREBERG, GEORGE A. PADGETT,
and ROY C. PAGE

INTRODUCTION

Investigators have long noted the involvement of the connective tissue
system in various diseases. Furthermore, some genetic and degenerative
diseases appear to involve only specific components of this system.[1, 2]
In spite of the recognition of its importance, few spontaneously occurring models of connective tissue disease have been available.

A heritable connective tissue disease of dogs and mink has been reported[3-6] that resembles the rare Ehlers-Danlos syndrome (ED-S) of
man.[2, 7-9] The human syndrome is recognized primarily by its clinical
manifestations and mode of inheritance. The important clinical criteria
for the ED-S of man include fragility of the skin, hyperextensibility of
the skin and joints, and laxity of the skin. Skin fragility has been considered to be the primary clinical manifestation of the disease.[10-12]
Because of the skin weakness, numerous lacerations have been recorded, especially in areas of the body vulnerable to injury, and healing has resulted in the formation of broad, thin scars. The ED-S of
man has an autosomal dominant mode of inheritance.[2, 9, 13]

Investigators have not been in full agreement concerning the morphologic change that occurs in the dermis of human patients, and alterations in both collagen and elastic fibers have been described.[2] Despite
this controversy, it is believed that alteration in the collagenous framework can best account for the skin fragility. Jansen[14] proposed that
the skin defect involved an abnormal "wickerwork" arrangement of
the collagen bundles in the dermis. The biochemical defect or defects
associated with the ED-S of man have not been characterized.

80

The heritable connective tissue disease of dogs and mink that resembles the ED-S of man has been studied in this laboratory. This report concerns the clinical, genetic, histopathologic, and biochemical results of these studies.

RESULTS

CLINICAL STUDIES

The major clinical change observed in the ED-S of dogs and mink was increased fragility of the skin. The fragility occurred throughout the skin in affected dogs of all ages, and was of sufficient magnitude that the skin could be easily ruptured. Skin tensile strength of affected dogs was only 4 percent of that of nonaffected dogs. Numerous naturally occurring deep lacerations of the skin, which healed and formed broad, thin scars were observed in both young and mature affected dogs. The skin fragility was more pronounced in young affected mink than in the adult affected mink, but skin fragility was detected in the adult mink in the thinner skinned areas, such as on the legs and feet. The tensile strength of the skin of affected mink was only approximately 8 percent that of control mink. Lacerations were noted in the skin of young affected kits; also abrasions of the head, shoulders, and back were more commonly seen than lacerations in the adult affected mink.

Hyperextensibility of the skin was recorded in all affected dogs and mink. Skin laxity was observed in affected dogs and mink and varied in severity among individual animals. The skin laxity was more pronounced in the adult affected dogs and mink, especially about the head, legs, and rump.

The skin of affected dogs appeared thin and blanched and the hair coat was often dry and fine in texture. No marked differences were noted in the appearance of the skin or pelage of affected mink.

Other clinical abnormalities noted in affected dogs included umbilical and inguinal hernias, luxated patellas, and rectal prolapses. These clinical changes were not observed in affected mink.

GENETIC STUDIES

The disease in both dogs and mink is genetically transmitted as an autosomal dominant trait. In studying the mode of inheritance, we found that both affected and nonaffected offspring were born from affected-dam-to-affected-sire matings, indicating that the disease is not recessive. Approximately 57 percent of the puppies and 51 percent of the kits from affected-sire-to-nonaffected-dam matings were affected, which also demonstrated a dominant mode of inheritance. The inheritance pattern was autosomal as both affected male and affected female offspring resulted from affected sire-to-nonaffected-dam matings in both dogs and mink. No significant difference was noted in the sex distribution of the disease in either affected dogs or mink.

Our breeding data indicated that the gene involved in the ED-S of dogs and mink is transmitted with complete penetrance. In dogs, nonaffected F_1 generation offspring from affected-dam-to-affected-sire matings did not produce affected F_2 generation offspring, suggesting that the affected dam and sire were heterozygotes. When one affected canine sire was mated to several nonaffected females, approximately 50 percent of the offspring were affected; this 1:1 ratio would be expected in heterozygote-to-normal-dog matings when the penetrance is complete. In addition, no generation skips were noted in the pedigrees of either ED-S dogs or mink, as nonaffected offspring from affected-to-affected or affected-to-nonaffected matings have never transmitted the defect to their F_2 progeny.

HISTOPATHOLOGIC STUDIES OF THE SKIN

Pronounced microscopic changes were present in the dermis of affected dogs and mink in all areas examined. The most severe alteration involved the collagenous component and included fragmentation and alteration in size and orientation of collagen bundles. Mature collagen bundles were fragmented and frayed rather than having the homogeneous appearance of the collagen bundles of the dermis of normal dogs and mink. Some collagen bundles were small in diameter (3 to 4 μ) and appeared as fibrillar strands, whereas other collagen bundles were larger than normal, measuring up to 40 μ in diameter. Collagen bundles were often haphazardly arranged and lacked the syncytoid interlacing arrangement seen in the normal dermis. Some collagen bundles in the dermis of affected animals were formed in a whorled arrangement. The degree of eosin staining varied in the collagen bundles in the dermis of affected animals; some collagen bundles were intensely stained, especially the larger bundles, whereas small bundles were only faintly stained.

Ground substance of the dermis was increased in approximately

30 percent of affected dogs and approximately 50 percent of the affected mink examined. This increase was pronounced in the papillary dermis and was characterized by deposition of metachromatic material located in small pools, or as small beads aligned on fine collagen fibers in the extracellular spaces.

No difference was noted in the appearance or in the apparent number of elastic or reticular fibers of affected dogs or mink as compared with their controls. The dermis of affected dogs was significantly thinner than that of control dogs, but no significant difference was noted in the dermal thickness of affected mink as compared with nonaffected mink. Little change was noted in the epidermis and adnexal structures of affected dog or mink skin.

BIOCHEMICAL STUDIES OF THE SKIN

Skin components, including water, collagen, elastin, and lipids were measured in young adult affected and control littermate dogs to determine whether differences could be found to explain the skin defect. A decrease in both the total skin collagen content (25.5 percent) and collagen per gram fat-free dry skin (13.4 percent) was found in the affected dogs as compared with the controls. The water content of the total skin was increased (2.3 percent), and the content per gram wet-weight skin was decreased (8.1 percent) in the affected dogs as compared with the controls. The lipid content was increased almost threefold in the affected dogs' skin as compared with the controls on a gram wet-weight basis. Little change was noted in the amount of elastin or soluble components of the skin other than gelatin or lipids.

Collagen was sequentially extracted in $1M$ NaCl and $1M$ acetic acid at $4°C$ from minced skin samples from affected and control littermate dogs. The amount of collagen, as determined by hydroxyproline content, was measured in the crude extracts both before and after dialysis. No significant difference was noted in the amount of extractable NaCl soluble collagen per gram wet-weight skin from affected dogs as compared with the controls. The amount of extractable acetic acid soluble collagen per gram wet-weight skin from affected dogs was increased up to 4 times that of the controls.

Crude nondialyzed, crude dialyzed, and purified $1M$ NaCl and $1M$ acetic acid extracted collagen from affected and control skins was electrophoresed on starch gel. No alteration in the alpha and beta band patterns, or in the migratory patterns, of other collagen fractions was noted. No abnormal bands were noted in the affected collagen from any of the samples. Amino acid analyses performed on affected and control-purified collagen revealed no significant differences.

DISCUSSION

CORRELATION OF SOME MAJOR FEATURES OF THE ED-S OF MAN, DOGS, AND MINK

Skin fragility is a consistent clinical finding in the ED-S of man and has been considered to be the most important clinical sign.[10-12] Rollhauser[15] found that the tensile strength of the skin from a 35-year-old man afflicted with the ED-S was only 20 percent that of normal human skin. The primary clinical feature of the disease in dogs and mink also is fragility of the skin. The tensile strength of affected dogs and affected mink was markedly reduced as compared with their controls. Skin hyperextensibility has been closely associated with the human syndrome[2, 9] and skin laxity has been reported.[10] Hyperextensibility of the skin, and, to a lesser extent, laxity of the skin, have been observed in affected dogs and mink. Human ED-S patients often have hyperextensible and luxated joints.[2, 9, 10] Hyperextensibility of the joints also has been noted in young affected puppies.

The genetic pattern of the ED-S of man is autosomal dominant.[2, 9, 13] The syndrome in dogs and mink is genetically transmitted as an autosomal dominant trait also.

Changes in both collagen and elastic fibers of the dermis have been described in human ED-S cases. The morphologic changes in the dermis of affected dogs and mink include primarily fragmentation and alteration in the size and orientation of collagen bundles. No remarkable change in the dermal elastic fibers has been observed in the affected dogs and mink.

The important clinical features, inheritance pattern, and histopathologic changes of the Ehlers-Danlos syndrome of man appear to be the same as in the heritable disease of dogs and mink. More intensive studies of these aspects of the disease in dogs and mink may shed additional light on the pathogenesis of the human ED-S.

CORRELATION OF FUNCTIONAL AND BIOCHEMICAL CHANGES

All hypotheses advanced concerning the basic defect involved in the Ehlers-Danlos syndrome must explain the clinical changes in skin strength. Other clinical signs closely associated with the ED-S, including skin hyperextensibility and laxity, joint hyperextensibility, and hernias, probably result from this same basic defect. Collagen is the primary component of the skin that contributes to its tensile strength,[15-17] and either quantitative or qualitative changes in collagen could result in a decrease in tensile strength. Quantitative changes include an absolute or a relative decrease in the number of collagen fibers per unit area. Qualitative changes involve alteration of collagen cross-linking, and include defects intrinsic to the collagen polypeptide or extrinsic factors other than collagen that might be important in a cross-linking phenomenon.

Some quantitative change was noted in the collagen and water content of the skin of affected dogs as compared with their normal littermate controls. Although these data were not expressed in terms of change per unit area of skin, the changes were not considered sufficient to explain the reduction in tensile strength by either an absolute decrease in the amount of collagen in the skin, or an increase in the water content of the skin affecting the relative distribution of the collagen fibers.

A more plausible explanation of the reduced tensile strength of the skin deals with a qualitative change affecting the cross-linking of collagen. Either an intramolecular or intermolecular collagen cross-linking defect can produce a decrease in tensile strength in tissues. Levene and Gross[18] reported that the decrease in tensile strength of tissues in lathyritic animals correlates with an increase in the amount of neutral NaCl soluble collagen in lathyritic tissues. Martin *et al.*[19] demonstrated that intramolecular cross-link formation in lathyritic collagen is defective. Because of the decrease in intramolecular cross-link formation in lathyrism, there is an increase in the amount of alpha collagen (single-stranded polypeptide chain), and a corresponding decrease in beta collagen (dimers of alpha collagen which are intramolecularly cross-linked).[20] An intramolecular cross-link, which is an aldol condensation product of δ-semialdehyde of a-amino adipic acid, has been identified.[21] We detected no significant increase in the amount of NaCl soluble collagen and no electrophoretic alteration in the alpha or beta components from the skin of ED-S dogs as compared with their normal littermate controls. These findings suggest that the defect is not one involving intramolecular collagen cross-linking.

Intermolecular collagen cross-links are believed to be even more important than intramolecular cross-links in imparting tensile strength to tissues.[22, 23] Bailey[23] showed that reduction of native tendon collagen resulted in increased tensile strength. Tanzer[24] and Bailey[23] indicated that a reducible, unstable, Schiff-base intermolecular cross-link existed in reconstituted and native collagen fibrils. They proposed that the Schiff-base-containing cross-link was intermediate in the formation of a stable covalent intermolecular collagen cross-link. Bailey and Peach[25] identified a reduced cross-link as hydroxylysino-norleucine, and they suggested that hydroxylysino-norleucine represented a stable covalent intermolecular cross-link of native collagen. As weak organic acids are believed to dissociate labile Schiff-bases of collagen fibrils, acid-soluble collagen probably represents collagen undergoing formation of this stable intermolecular cross-linkage.

We have found as much as a fourfold increase in the amount of acetic-acid-soluble collagen extractable from the skin of ED-S dogs, as compared with normal littermate controls. Similar findings of increased recovery of acetic-acid-soluble collagen from the skin of human ED-S patients has been reported.[26] We believe that the skin changes of decreased tensile strength and increased acid-soluble collagen content of ED-S dogs may result from a specific defect in stable intermolecular collagen cross-link formation. Intramolecular cross-link formation does not appear to be altered in the ED-S dogs, suggesting that intramolecular and intermolecular collagen cross-links are formed, at least partially, by separate processes.

CORRELATION OF THE BIOCHEMICAL CHANGE AND GENETIC MECHANISMS

Most defined genetic diseases of a simple Mendelian nature involve a specific alteration in a structural protein or a defect in an enzyme system. As the ED-S is a genetically transmitted disease, such a working hypothesis should be valid for the molecular defect in this disease also.

In this report, we have suggested that the defect in the ED-S may involve an inability of collagen to form stable intermolecular cross-links. Such a defect could arise either from an intrinsic collagen abnormality or from an extrinsic enzyme defect affecting collagen-to-collagen cross-linking, or cross-linking of collagen by noncollagenous factors.

Substitution of amino acids whose side chains are reactive in the formation of intermolecular cross-links, such as lysine and hydroxylysine that are important in the cross-link identified by Bailey and Peach,[25] would constitute an intrinsic error of collagen formation affecting the capacity to cross-link. No alteration in electrophoretic patterns or amino acid analyses was detected in the ED-S dog collagen; however, a structural collagen defect cannot be ruled out on the basis of these results alone.

Although it has been suggested that an amino oxidase is necessary for an oxidative deamination step[27] in the formation of intramolecular collagen cross-links, little is known concerning the role of extrinsic factors in intermolecular cross-link formation. Some workers have proposed that insoluble collagen fibrils form spontaneously *in vitro* and *in vivo*, independent of any extrinsic factors.[28] On the other hand, Tsurufuji and Ogata[29] concluded that formation of insoluble collagen in minced skin samples was dependent on normal cellular activity, suggesting that vital extrinsic processes are necessary for maturation of collagen.

Extrinsic factors might influence intermolecular cross-link formation at one of at least three possible sites. One site involves a potential enzyme system responsible for the extraribosomal hydroxylation of lysine residues in the collagen polypeptide. The mechanism of hydroxylation of lysine is believed to occur in a manner similar to the hydroxylation of proline.[30] The conversion of proline to hydroxyproline is catalyzed by an enzyme, proline hydroxylase.[31] As hydroxylysine forms a part of the hydroxylysino-norleucine cross-link, a defect in the hydroxylating enzyme might result in defective cross-link formation.

A second possible site is in the process of formation of the hydroxylysino-norleucine cross-link. The formation of this cross-link could involve several enzymatically controlled stages, including the formation of aldehyde-rich hydroxylysine or lysine residues, formation of a Schiff-base intermediate, and the subsequent reduction of Schiff-base to form a stable covalent cross-link.

A third possibility would be the stabilization of collagen by cross-links formed with extracollagenous components of the skin. Several workers[32, 33] have presented evidence that collagen aggregates are stabilized by forming complexes with mucopolysaccharides or mucoproteins. The chemical nature and mechanism of formation of these cross-links have not been defined.

A defect in a hydroxylysino-norleucine intermolecular cross-link at an intermediate Schiff-base formation stage offers the most interesting possibility as the site of the defect in the ED-S and would explain the increase in acid-soluble collagen and the decrease in skin tensile strength.

Studies are presently under way to delineate further the biochemical defect in the ED-S of dogs and mink. It is hoped that the ED-S of dogs and mink may provide additional information concerning the biochemical abnormality of the disease in man, and may help delineate the mechanisms involved in collagen cross-linking and maturation.

SUMMARY

A connective tissue disease of dogs and mink that resembles the Ehlers-Danlos syndrome (ED-S) of man was investigated. Prominent clinical signs noted in affected dogs and mink included fragility, hyperextensibility, and laxity of the skin. The inheritance pattern of the disease in dogs and mink is autosomal dominant. The primary histopathologic changes included fragmentation and alteration of size and orientation of collagen bundles of the dermis. The clinical changes, mode of inheritance, and histopathologic changes in the skin for the disease in dogs and mink appear identical to those described for the ED-S of man.

The important biochemical change found in the ED-S dogs was a significant elevation in the amount of acid-soluble collagen that could be extracted from the skin. Recent evidence by other workers indicates that intermolecular cross-links give stability to collagen, and that acid-labile, Schiff-base intermediates are involved in the formation of this stable bond. We suggest that the increase in acid-soluble collagen in the skin may represent a specific alteration in an intermolecular cross-link. This cross-link may be involved in maturation of collagen from an acid-soluble to a more insoluble state.

This study was supported in part by NIH grants 5-SO1-5465, 1-F3-FR-39, 627-01, GRSF 171 Proj., and 03174.

REFERENCES

1. Klemperer, P., A. D. Pollack, and G. Baehr. 1942. Diffuse collagen disease. J. Amer. Med. Ass. 119:331–332.
2. McKusick, V. A. 1966. Heritable disorders of connective tissue. Third edition. C. V. Mosby Co., St. Louis.
3. Hegreberg, G. A., G. A. Padgett, R. L. Ott, and J. B. Henson. A heritable connective tissue disease of dogs and mink resembling the Ehlers-Danlos syndrome of man. I. Clinical changes and skin tensile strength properties. J. Invest. Derm. In press.
4. Hegreberg, G. A., G. A. Padgett, J. R. Gorham, and J. B. Henson. 1969. A heritable connective tissue disease of dogs and mink resembling the Ehlers-Danlos syndrome of man. II. Mode of inheritance. J. Heredity 60:249–254.

5. Hegreberg, G. A., G. A. Padgett, and J. B. Henson. A heritable connective tissue disease of dogs and mink resembling the Ehlers-Danlos syndrome of man. III. Histopathologic changes of the skin. Archives of Pathology. In press.

6. Hegreberg, G. A., and R. C. Page. 1969. A heritable disease of dogs involving collagen maturation. Fed. Proc. 28:622.

7. Ehlers, E. 1901. Cutis laxa, Neigung zu Harmorrhagien in der Haut, Lockerung mehrerer Artikulationen. Derm. Zachr. 8:173–174.

8. Danlos, M. 1908. Un cas de cutis laxa avec tumeurs par contusion chronique des coudes et des genoux (xanthome juvenile pseudo-diabetique de M. M. Hallopeau et Mace de Lepinay). Bull. Soc. Franc. Derm. Syph. 19:70–72.

9. Johnson, S. A. M., and J. F. Falls. 1949. Ehlers-Danlos syndrome: a clinical and genetic study. Arch. Dermatol. Syph. 60:82–105.

10. Summer, G. K. 1956. The Ehlers-Danlos syndrome, a review of the literature and report of a case with a subgaleal hematoma and Bell's palsy. J. Dis. Child. 91:419–428.

11. Weber, F. P., and J. K. Aitken. 1938. Nature of the subcutaneous spherules in some cases of Ehlers-Danlos syndrome. Lancet 1:198–199.

12. Ronchese, F. 1936. Dermatorrhexis with dermatochalasia and arthrochalasis (the so-called Ehlers-Danlos syndrome). Amer. J. Dis. Child. 31:1404–1414.

13. Jansen, L. H. 1955. Le mode de transmission de la maladie d'Ehlers-Danlos. J. Genet. Hum. 4:204–218.

14. Jansen, L. H. 1955. The structure of the connective tissue, an explanation of the symptoms of the Ehlers-Danlos syndrome. Dermatologica 110:108–120.

15. Rollhauser, H. 1950. Die zugfestigkeit der menschlichen haut. Gegenbaur. Morph. Jahrb. 90:249–261.

16. Fry, P., M. L. R. Harkness, and R. D. Harkness. 1964. Mechanical properties of the collagenous framework of skin in rats of different ages. Amer. J. Physiol. 206:1425–1429.

17. Varadi, D. P., and D. A. Hall. 1965. Cutaneous elastin in Ehlers-Danlos syndrome. Nature 208:1224–1225.

18. Levene, C. I., and J. Gross. 1959. Alterations in the state of molecular aggregation of collagen induced in chick embryos by B-amino propionitrile (Lathyrus factor). J. Exp. Med. 110:771–791.

19. Martin, G. R., J. Gross, K. A. Piez, and M. S. Lewis. 1961. On the intramolecular cross-linking of collagen in lathyritic rats. Biochim. Biophys. Acta 53:599–601.

20. Page, R. C., and E. P. Benditt. 1967. A molecular defect in lathyritic collagen. Proc. Soc. Exp. Biol. Med. 124:459–465.

21. Bornstein, P., A. H. Kang, and K. A. Piez. 1966. The nature and location of intramolecular cross-links of collagen. Proc. Nat. Acad. Sci. 55:417–424.

22. Piez, K. A. 1968. Cross-linking of collagen and elastin. Ann. Rev. Biochem. 37:547–570.

23. Bailey, A. J. 1968. Intermediate labile intermolecular crosslinks in collagen fibres. Biochim. Biophys. Acta 160:447–453.

24. Tanzer, M. L. 1968. Intermolecular cross-links in reconstituted collagen fibrils. J. Biol. Chem. 243:4045–4054.

25. Bailey, A. J., and C. M. Peach. 1968. Isolation and structural identification of a labile intermolecular crosslink in collagen. Biochem. Biophys. Res. Communs. 33:812–819.

26. Harris, E. D., Jr., and A. Sjoerdsma. 1966. Collagen profiles in various clinical conditions. Lancet 2:707–711.
27. Miller, E. J., S. R. Pinnell, G. R. Martin, and E. Schiffmann. 1967. Investigation of the nature of the intermediates involved in desmosine biosynthesis. Biochem. Biophys. Res. Communs. 26:132–137.
28. Gross, J. 1968. Studies on the formation of collagen. III. Time-dependent solubility changes of collagen in vitro. J. Exp. Med. 108:215–226.
29. Tsurufuji, S., and Y. Ogata. 1965. Biosynthesis of collagen in skin minces in relation to the mechanism of the formation of insoluble collagen. Biochim. Biophys. Acta 104:193–199.
30. Sinex, F. M., D. D. Van Slyke, and D. R. Christman. 1966. The source and state of the hydroxylysine of collagen. II. Failure of free hydroxylysine to serve as a source of the hydroxylysine or lysine of collagen. J. Biol. Chem. 234:918–921.
31. Hutton, J. J., and S. Udenfriend. 1966. Soluble collagen proline hydroxylase and its substrates in several animal tissues. Proc. Nat. Acad. Sci. 56:198–202.
32. Wood, G. C. 1962. The heterogeneity of collagen solutions and its effect on fibril formation. Biochem. J. 84:429–435.
33. Partington, F. R., and G. C. Wood. 1963. The role of non-collagen components in the mechanical behaviour of tendon fibres. Biochim. Biophys. Acta 69:485–495.

Psychological Factors in Comparative Biomedical Research

ROBERT ADER

The single point I wish to make is that psychological factors are capable of influencing an individual's susceptibility to disease. In comparative biomedical research, much attention has been directed toward the search for specific species and strains of animals that, by virtue of their genetic uniformity or special anatomic or physiologic endowments, are particularly suited for the experimental reproduction and study of specific disease processes that exist in man.[18, 20] The history of experimental and clinical medicine attests to the success and value of this approach in uncovering agents of disease. Coincidentally, the precise parametric analysis of the action of various pathogenic agents, perhaps by its very nature, acts to reduce the influence of factors that contribute to the development of disease as it evolves in the real world. It may be presumed that such factors have not been completely eliminated since variability remains one of the outstanding features of all biologic research. Despite passing references to the role of psychological variables in the etiology of organic disease, by and large this proposition has been experimentally neglected.

Under normal circumstances individuals are not exposed to pathogens via a route, or in a dose, that will unconditionally elicit disease. It is commonplace to observe that, in a population of individuals all of whom are exposed to some common pathogen, only some individuals succumb to the disease. Why? What factors contribute to the development of manifest disease in a way that makes some individuals more susceptible than others? Such individual differences may be ascribed to biological differences or innate variability, but these are descriptions, not explanations, and serve only to point up our lack of knowledge concerning the multitude and complexity of the variables and interactions of variables that may be inferred to exist.

Attempts to achieve uniformity (e.g., through inbreeding, germ-free environments, environmental constancy, etc.) have never been totally successful in eliminating individual differences. And to the extent that we are interested in analogies to the human condition this is a desirable state of affairs. Variability cannot be denied; it can be ignored or it can stimulate hypotheses and further experimentation. I, of course, advocate the latter. Indeed, there is ample evidence to indicate that individual differences are amenable to analysis and are, at least in part, attributable to the individual's past history. By past history I mean anything and everything that has contributed to the characteristics of the individual. Specifically, I am referring to the interaction of genic factors, prenatal influences, early life experiences, and the social environment in which a person has lived. No one would question the role of heredity in determining certain biologic limits, or that the behaving individual represents a complex interaction of genetically and environmentally determined characteristics. Moreover, with respect to the effects of early life experiences, for example, it can be argued, that the phenotypic expression of an individual's genetic endowment can be—and is quite likely to be—influenced by the environment, or by experiences occurring early in life when growth and development are particularly rapid and dramatic. I refer not only to behavior, but also to a variety of other biologic characteristics including rate of growth and ultimate size, anatomic and physiologic characteristics, and resistance to disease.[17] I would also argue that the interaction between genes in determining the biologic limits of certain characteristics is no more complex than is the interaction of environmental events that go into making the individual what he is.

There is nothing in a psychosomatic orientation to imply that psychological factors are sufficient in themselves to cause disease. Psychological factors can exert their effects in a variety of ways—e.g., by modifying neuroendocrine status chronically via early life experiences or acutely by virtue of prevailing psychosocial conditions, by altering psychophysiological thresholds, or by affecting responses and adaptation to the disease. However, psychological variables represent only one possible link in the chain of events that determines susceptibility to disease. Our own research has been concerned with the influence of early life experiences on the development and subsequent behavior of animal subjects and their susceptibility to disease. These data, which have been reviewed elsewhere,[19] provide ample evidence that experiences, particularly those that occur during early life, are capable of altering the psychophysiological functioning of the individual. We do not yet know the mechanisms whereby psychological experiences are translated into altered physiological states, or how such psychophysiological changes interact with the responses unconditionally elicited by the superimposition of potentially pathogenic stimuli to modify susceptibility to disease. Still, the many demonstrations of psychosomatic phenomena in animals warrants a more direct consideration of the role of experiential factors in the study of animal models of disease.

Rather than provide a survey of the variety of disease processes that have been found to be influenced by a variety of psychosocial variables, I shall illustrate the effects of specific experiential variables on a single pathogenic process—experimentally induced gastric lesions in the rat. There are a number of techniques for eliciting gastrointestinal lesions in animals (burning, drug administration, pyloric ligation, brain stimulation, nutritional manipulations). Physical immobilization is particularly effective; it is simple to administer, gastrointestinal changes are induced rapidly, and it does not involve surgery or the administration of drugs. These advantages make the immobilization technique particularly well suited for the study of gastrointestinal physiology, the effects of physiologic and pharmacologic manipulations, and the role of psychological factors in the development of ulcers. With this technique the lesions are confined, almost exclusively, to the lower, glandular, acid-secreting portion of the stomach. Systematic studies of the factors involved in the production of gastric lesions by immobilization have shown that the development of acute lesions is inversely related to the degree, and directly related to the duration, of restraint, and that food deprivation prior to immobilization exerts a deleterious effect upon the development of the lesions and upon recovery of the stomach.[1,13–15] Further, these are strain and sex differences in erosion susceptibility,[11] and rats can be selectively bred for susceptibility.[26, 27]

Several studies in man have shown that patients with duodenal ulcers have higher plasma pepsinogen levels than people without gastrointestinal disturbances.[21] Similarly, rats that are immobilized and subsequently found to have gastric erosions have higher plasma pepsinogen levels than immobilized animals that do not develop erosions.[11] Further, studies in man have shown that plasma pepsinogen level is a predictor of susceptibility to duodenal ulcer.[30, 35] As these investigators point out, however, the relationship between pepsinogen level and susceptibility to duodenal ulcer is not a simple physiological one. Pepsinogen level, taken as a measure of gastric secretory potential, represents a biologic vulnerability or predisposition that interacts with the individual's psychologic capacity to adapt to changes in his psychosocial environment, and it is this interaction that actually determines the individual's liability to the disease. The likelihood of developing duodenal ulcer would be maximal in an individual who is biologically predisposed (i.e., has a relatively high plasma pepsinogen level), and whose personality structure is such that he is unable to cope with environmental events that he perceives as being "stressful" or threatening. It is the complexity of such psychophysiological interactions, and the procedural (and ethical) difficulties that would be met in manipulating and testing the effects of the several factors that are involved, that prompted our own attempts to examine the effects of psychological factors on susceptibility to ulceration in the rat.

Having found that rats with gastric erosions have higher plasma pepsinogen levels than do animals without erosions,[10] and that the distribution of plasma pepsinogen levels in the rat is the same as that in man,[2] we selected rats with basal pepsinogen levels that fell within the highest and the lowest 15 percent segments of the distribution. Subsequent immobilization of these extreme populations indicated that the rats with high pepsinogen levels are more likely to develop gastric erosions than are animals with low pepsinogen levels.[1] However, this statistically predictive relationship was observed only when the rats were subjected to a 6-hour period of immobilization, which is minimally conducive to the development of lesions. Longer periods of restraint produced lesions in a greater percentage of the animals, and there was no predictive relationship between pepsinogen level and susceptibility. The interaction between pepsinogen level and the duration of immobilization thus represents an interesting example of the apparent fact that, if potentially pathogenic stimulation is sufficiently intense to elicit pathology unconditionally, there is little latitude for the identification of underlying biologic (or psychologic) processes that may be relevant under conditions that prevail in a natural environment.

Not all animals with high pepsinogen levels developed erosions when subjected to immobilization, nor did a low pepsinogen level uniformly afford protection against the development of lesions. It was concluded, therefore, that a high pepsinogen level is neither necessary nor sufficient for the development of immobilization-induced gastric erosions. Nonetheless, on a statistical basis, it remains true that a high pepsinogen level can be taken as a biologic indicator of heightened susceptibility to gastric erosions in a manner that appears to be analagous to the relationship between pepsinogen level and duodenal ulcer in man.

Considering the extent of unaccounted for variability, it can (and should) be asked why all animals with a similar biologic predisposition (i.e., high pepsinogen levels) did not develop gastric erosions when all animals were subjected to the same stimulus conditions. If such results were observed in a population of humans it could be argued that what is "stressful" for one individual is not necessarily "stressful" for another, that different individuals perceive and respond differently to the same stimuli. On the assumption that the same argument holds for the data derived from animals, an attempt was made to vary the manner in which rats might perceive and respond to a constant stimulus.[3]

Since gastric erosions are so easily induced by physical restraint, its opposite, free activity, suggested itself as a behavior that might be related to erosion susceptibility. Given the advantage of the characteristic circadian rhythm in activity, one group of rats was immobilized as they approached the peak, and a second group as they approached the trough in their individually determined cycles of activity. Since an increase in "stress" (the duration of immobilization) results in an increase in the percentage of rats that develop erosions, it was hypothesized that animals immobilized during the active period of their cycle should perceive and respond to the restraint as if it were more "stressful" than should animals immobilized during a trough in the daily rhythm. The hypothesis was confirmed; all the animals that developed erosions were from among those restrained during the peak of their activity cycle. These results were repeated in a subsequent study[6] in which an attempt was made to delineate the potential effects of other contributing factors. Although these studies have not eliminated all possible physiologic mediators of this 24-hour susceptibility rhythm, the hypothesis relating gastric erosion susceptibility to psychological factors (i.e., the individual's "perception" of the environmental conditions) remains tenable. At the very least, these data clearly demonstrate that the existing psychophysiological state of the individual is a meaningful factor in determining responses to superimposed stimulation.

The influence of psychological factors in the pathogenesis of experimentally produced gastric lesions is most apparent in studies where the past history of the animals has been experimentally manipulated or in more acute studies in which psychological factors act to modify the effects of other ulcerogenic stimuli. For example, Csalay, Frenkl, and Hegyváry[16] reported that rats subjected to prolonged visual, auditory, and electric shock stimulation will develop "neurogenic hypertension," but that this stimulation is ineffective in inducing gastric lesions. However, this same stimulation will increase the ulcerogenic effects of serotonin. Also, such stimulation administered to rats fed a methionine-deficient diet results in the development of rumenal and glandular ulceration although neither the stimulation nor the methionine-deficient diet alone is ulcerogenic. Similarly, Rosenberg[22] subjected rats housed in a cold (5–8°C) room to continuous electric shock for a period of 8 hours. Fifty percent of the animals developed glandular lesions. Twelve hours of such stimulation increased the incidence of lesions to 82 percent. In contrast, cold or electric shock stimulation alone produced lesions in only 10 to 15 percent of the animals.

A number of studies have been made to determine the effects of fear or anxiety on the development of gastric erosions. In an experiment by Weisz,[33] rats were group-housed in cages with grid floors for 30 days and kept on a 47-hour deprivation schedule while being subjected to light-shock stimulation. For half the trials the light stimulus and electric shock were presented simultaneously; in the remaining trials only the light was presented. One control group received the same stimulation but was allowed free access to food and water; another control group experienced only the deprivation. Fear plus deprivation resulted in a significantly increased susceptibility to gastric erosions, compared with the conditions of fear without deprivation, or of deprivation alone.

Sawrey[23] subjected rats to regular (predictable) or irregular (unpredictable) schedules of electric shock stimulation. The animals were deprived of food and water for 22 of each 24 hours and subjected to the random presentation of a light and buzzer. The "regular" group received electric shock after each presentation of the light stimulus while the "irregular" group received electric shock following half the light and half the buzzer presentations. Thus, for the "regular" group, the occurrence of shock was predictable, whereas shock was unpredictable for the "irregular" group. Although the number of light and buzzer presentations and the number of electric shocks were equal, the rats subjected to the unpredictable schedule developed significantly more gastric lesions than did the animals that experienced the predictable shock schedule. Presumably, the regular schedule of shock would have elicited a fear response only when the light was presented while the unpredictability of the irregular shock schedule would have elicited a fear response upon presentation of the light and the buzzer. Therefore, the "irregular" or unpredictable shock group would have experienced twice as many fear stimuli as the "regular" group. In a subsequent study Sawrey and Sawrey[24] found that prior experience with predictable or unpredictable electric shock increased resistance to ulceration and eliminated the difference in susceptibility between deprived animals experiencing the predictable and unpredictable shock schedules.

Sawrey and Sawrey[25] also studied the effect of fear on susceptibility to immobilization-induced gastric lesions. Rats were given 20 or 80 trials consisting of the presentation of a light stimulus which was associated with electric shock half the time. The animals were then immobilized for 48 hours during which the light stimulus occurred repeatedly but shock was not applied. Although 20 fear-conditioning trials did not influence the development of gastric lesions, 80 fear-conditioning trials significantly increased susceptibility relative to control animals that received only the light stimuli prior to immobilization.

In order to study the effects of the ability to cope with a "stressful" situation, Weiss[32] placed rats in small restraining cages for a period of 21 hours. One group of animals was presented with a conditioned stimulus for 10 seconds followed by an electric shock which could be avoided if the animal used its nose or paw to touch a plate directly in front of its cage. A response after the onset of shock served to terminate shock. A second group served as a yoked control, receiving shock at the same frequency and duration as the animals that were capable of avoiding it; a third group was restrained but received no electric shocks. The animals were sacrificed after a subsequent 12-hour period of deprivation. There was some tendency for the animals that were capable of avoiding or escaping shock to be more susceptible to ulceration than the nonshocked animals, but this difference was not consistent. However, the yoked animals—those that were unable to control the presentation or termination of shock—were significantly more susceptible to gastric lesions than both the "avoidance" and control groups.

Studies in which the redundant history of animals has been manipulated prior to the induction of gastric lesions also demonstrate the potential effect of experiential factors. One of the easiest psychosocial variables to manipulate is the manner in which laboratory animals are housed. Numerous studies have shown that group-caged and individually caged rats differ in response to a variety of pathogenic agents.

With respect to immobilization-induced gastric erosions, rats that are housed in groups from weaning until maturity are significantly more susceptible to lesions than are animals housed individually.[4, 28, 29] No behavioral or physiological analysis of this phenomenon has been proposed, simply because a systematic study of the pathogenesis of gastric lesions has not been undertaken with differentially housed animals. Group-housed rats are less emotionally reactive than rats caged individually, but no consistent relationship between emotional reactivity and lesion susceptibility in the rat has been demonstrated.[7]

Experimentally manipulating the early life experiences of the laboratory rat has profound effects on its development and on subsequent behavioral and physiological processes. Data have been accumulating to show that experiences occurring during early life are also capable of influencing the organism's response to organic disease. The majority of studies of early experience have involved direct stimulation of the infant animal by the experimenter, and the most common form of such stimulation consisting of merely handling the animal. The term "handling" includes a range from holding the animal to simply removing it from the nesting cage and placing it into a new environment for a short period of time.

Again, with respect to immobilization-induced gastric erosions, Weininger[31] found that handled animals were less susceptible to lesions than nonhandled littermate controls. The individually housed animals received 10 minutes of handling each day for the 21 days immediately after weaning (21 days) and were later subjected to 48 hours of immobilization. Winokur, Stern, and Taylor[34] obtained similar results in a study in which rats handled for 5 or 10 minutes each day during the postweaning period were subsequently group-housed.

Ader[4] subjected rats to 3 minutes of handling or electric shock stimulation once each day during the first 21 days of life. After weaning, half the stimulated groups and an unmanipulated control group were group-housed and half were caged individually. At maturity all animals were subjected to 18 hours of immobilization. Among the group-housed animals, handling reduced susceptibility to gastric erosions relative to electric shock stimulation; animals that had experienced electric shock were more susceptible than controls. In the individually housed population, handled animals were significantly less susceptible to lesions than were shocked or unmanipulated animals that did not differ from each other. These results, then, are in accord with the above findings even though there were differences in the strain and age of the animals used, in the type and duration of handling, in the age when the animals were handled, and in the type and duration of immobilization.

One further illustration goes back even further in experiential history. Instead of handling infant rats Ader and Plaut[12] subjected pregnant rats to daily handling throughout the period of gestation. To eliminate potential postnatal maternal influences, the offspring of handled and nonhandled females were fostered to nonhandled females at parturition. Although the offspring were group-housed after weaning, they were individually housed for a period of time before immobilization. In this study the female offspring of handled animals were more susceptible than control animals to gastric erosions; there were no differences between the male groups. In a second experiment half the offspring of handled and nonhandled females were group-housed and half were caged individually. There were no differences in susceptibility to lesions between the group-housed animals, but among the individually housed animals, the male and female offspring of those handled throughout pregnancy were more susceptible than the controls to gastric erosions.

The pathogenesis of gastric lesions in animals is extremely complex, but even from the limited sample of the literature cited here, it is evident that psychological factors can contribute to this process. Psychological factors do not cause ulceration; they exert their effects through interaction with environmental stimuli or with a biologic predisposition. Therefore, the study of the experiential factors influencing the development of experimentally induced gastric lesions in animals is an appropriate model for the study of psychosomatic processes. The reason it is a good model is not that one can generalize to man on the basis of the similarity of the disease to ulcers in man, or on the basis of common physiological mechanisms; it is a good model because it represents a pathologic process to which the animal is liable— a process that is sensitive to the interaction between psychological and physiological influences in a manner analogous to that hypothesized to exist in man.

I chose experimentally induced gastric lesions to illustrate the role of psychological factors in the pathogenesis of organic disease because the literature contains examples of the control of variability afforded by a knowledge of the psychophysiological state of the animal when it is exposed to potentially pathogenic stimulation, and of the effects of prenatal, early experiential, and current social factors on susceptibility. These data can be generalized. Although the range of experiential factors is not as great, there are numerous other examples of the effects of psychological variables on susceptibility to a variety of pathogenic agents. More extensive reviews of these data have been published elsewhere.[5, 8, 9, 19] As might be expected, a given experimental manipulation does not always produce an altered susceptibility to disease. That is, differential housing does not influence the response to *all* pathogenic stimuli, nor are group-housed animals *always* more (or less) susceptible than individually housed animals when differences in susceptibility are observed. Similarly, handling or any other form of stimulation to which infant animals may be exposed does not *always* effect a change in disease susceptibility, nor are experimentally manipulated animals *always* more resistant than controls. These facts, however, do not detract from the demonstrated influence of psychological factors on susceptibility to disease. They indicate only that we have yet to learn how experiences are translated into altered physiological states, and how these psychophysiological states interact with the responses unconditionally elicited by the superimposition of specific pathogenic stimuli to influence the development of manifest disease.

I think that most researchers, by intuition or by direct clinical or research experience, are coming to acknowledge the role of psychological factors in the etiology of disease in man. It has been my aim here to show that the role of psychological factors in the pathogenesis of organic disease can be experimentally manipulated and documented in animals. This leads me to conclude that a complete animal model of disease, as it occurs under natural conditions, must take into consideration the psychological and social background of the experimental subjects.

Preparation of this paper and the author's research were supported by United States Public Health Service Grants 1 KO5 MHO6318, MHO3655, and MH16741 from the National Institute of Mental Health.

REFERENCES

1. Ader, R. 1963a. Plasma pepsinogen level as a predictor of susceptibility to gastric erosions in the rat. Psychosom. Med. 25:221–232.
2. Ader, R. 1963b. Plasma pepsinogen level in rat and man. Psychosom. Med. 25:218–220.
3. Ader, R. 1964. Gastric erosions in the rat: Effects of immobilization at different points in the activity cycle. Science 145:406–407.
4. Ader, R. 1965. Effects of early experience and differential housing on behavior and susceptibility to gastric erosions in the rat. J. Comp. Physiol. Psychol. 60:233–238.

5. Ader, R. 1966. Early experience and adaptation to stress. Res. Publ. Ass. Res. Nerv. Ment. Dis. 43:292–306.

6. Ader, R. 1967a. Behavioral and physiological rhythms and the development of gastric erosions in the rat. Psychosom. Med. 29:345–353.

7. Ader, R. 1967b. Emotional reactivity and susceptibility to gastric erosions. Psychol. Rep. 20:1188–1190.

8. Ader, R. 1967c. The influence of psychological factors on disease susceptibility in animals. In M. L. Conalty (Ed.) The husbandry of laboratory animals. Academic Press: London.

9. Ader, R. 1970. Experimentally induced gastric lesions: Results and implications of studies in animals. Adv. Psychosom. Med. (In press).

10. Ader, R. 1970. The effects of early life experiences on developmental processes and susceptibility to disease in animals. In J. P. Hill [Ed.] Minnesota Symposium on Child Psychology 4.

11. Ader, R., C. C. Beels, and R. Tatum. 1960. Blood pepsinogen and gastric erosions in the rat. Psychosom. Med. 22:1–12.

12. Ader, R., and S. M. Plaut. 1968. Effects of prenatal maternal handling on offspring emotionality, plasma corticosterone levels, and susceptibility to gastric erosions. Psychosom. Med. 30:277–286.

13. Bonfils, S., G. Liefooghe, X. Gellé, M. Dubrasquet, and A. Lambling. 1960. "Ulcère" expérimental de contrainte du rat blanc: III. Mise en évidence et analyse du rôle de certains facteurs psychologiques. Rev. Franc. Etudes Clin. Biol. 5:571–581.

14. Bonfils, S., G. Rossi, G. Liefooghe, and A. Lambling. 1959. "Ulcère" expérimental du rat blanc: I. Methodes. Fréquence des lesions. Modifications par certains procedes techniques et pharmacodynamiques. Rev. Franc. Etudes Clin. Biol. 4:146–150.

15. Brodie, D., and H. M. Hanson. 1960. A study of the factors involved in the production of gastric ulcers by the restraint technique. Gastroenterology 38:353–360.

16. Csalay, L., R. Frenkl, and Cs. Hegyváry. 1962. Ulcerogenic action of chronic neurogenic stimulation in the rat. Acta Physiol. 22:81–87.

17. Dubos, R., D. Savage, and R. Schaedler. 1966. Biological Freudianism: Lasting effects of early environmental influences. Pediatrics 38:789–800.

18. Frenkel, J. K. 1969. Choice of animal models for the study of disease processes in man: Introduction. Fed. Proc. 28:160–161.

19. Friedman, S. B., L. A. Glasgow, and R. Ader. Psychosocial factors modifying host resistance to experimental infections. Ann. N.Y. Acad. Sci. 164:381–392.

20. Jones, T. C. 1969. Mammalian and avian models of disease in man. Fed. Proc. 28:162–169.

21. Mirsky, I. A. 1958. Physiologic, psychologic, and social determinants in the etiology of duodenal ulcer. Am. J. Digest. Dis. 3:285–313.

22. Rosenberg, A. 1967. Production of gastric lesions in rats by combined cold and electrostress. Am. J. Digest. Dis. 12:1140–1148.

23. Sawrey, W. L. 1961. Conditioned responses of fear in relationship to ulceration. J. Comp. Physiol. Psychol. 54:347–348.

24. Sawrey, W. L., and J. M. Sawrey. 1963. Fear conditioning and resistance to ulceration. J. Comp. Physiol. Psychol. 56:821–823.

25. Sawrey, W. L., and J. M. Sawrey. 1964. Conditioned fear and restraint in ulceration. J. Comp. Physiol. Psychol. 57:150–151.
26. Sines, J. O. 1959. Selective breeding for development of stomach lesions following stress in the rat. J. Comp. Physiol. Psychol. 52:615–617.
27. Sines, J. O. 1961. Behavioural correlates of genetically enhanced susceptibility to stomach lesion development. J. Psychosom. Res. 5:120–126.
28. Sines, J. O. 1965. Pre-stress sensory input as a non-pharmacologic method for controlling restraint-ulcer susceptibility. J. Psychosom. Res. 8:399–403.
29. Stern, J. A., G. Winokur, A. Eisenstein, R. Taylor, and M. Sly. 1960. The effect of group vs. individual housing on behaviour and phsyiological responses to stress in the albino rat. J. Psychosom. Res. 4:185–190.
30. Weiner, H., M. Thaler, M. F. Reiser, and I. A. Mirsky. 1957. Etiology of duodenal ulcer: I. Relation of specific psychological characteristics to rate of gastric secretion (serum pepsinogen). Psychosom. Med. 19:1–10.
31. Weininger, O. 1956. The effects of early experience on behavior and growth characteristics. J. Comp. Physiol. Psychol. 49:1–9.
32. Weiss, J. M. 1968. Effects of coping responses on stress. J. Comp. Physiol. Psychol. 65:251–260.
33. Weisz, J. D. 1957. The etiology of experimental gastric ulceration. Psychosom. Med. 19:61–73.
34. Winokur, G., J. A. Stern, and R. Taylor. 1959. Early handling and group housing: Effect on development and response to stress in the rat. J. Psychosom. Res. 4:1–4.
35. Yessler, P. G., M. Reiser, and D. McK. Rioch. 1959. Etiology of duodenal ulcer: II. Serum pepsinogen and peptic ulcer in inductees. J.A.M.A. 169:451–456.

The Use of Amphibians
in Biomedical Research

GEORGE W. NACE

INTRODUCTION

Are there any among you who are not plagued each spring by hordes of boys with frogs in hand, anxious to tell you about frogs, to ask you about them, or to request your assistance in demonstrating their viscera? If this were the American Medical Association instead of the American Veterinary Medical Association, I would also ask if there are any among you who saw the viscera of any animal before you had become acquainted with those of the frog. What other justification is needed for the use of amphibians in biomedical research? What can be more significant than for the teacher-investigator-veterinarian to be the vehicle for bringing young, inquiring minds to biological research?

In addition to this ontogenic interest, frogs also have remained in the limelight throughout the phylogeny of biological and medical research. Galvani first demonstrated that scraping metal against metal produces electricity by using the twitch of a frog leg to indicate the discharge; and Volta then elaborated on this phenomenon to produce the first battery.[1]

Perhaps it is because the amphibian is so familiar that we have been so cavalier in our attention to it as an experimental animal and, indeed, know so little about its biology and pathology. How many investigators demand extreme accuracy from their measuring instruments, and yet are willing to use animals that are not standardized as to nutrition, health, genetics, or phase of reproductive or seasonal physiological cycles?

In my efforts to examine protein ontogeny in amphibians,[2] I eventually encountered serious difficulties because it was impossible to determine whether the qualitative variations were a result of an ontogenic sequence or of genetic variability among the test animals. I must not be alone in having encountered such problems, for 277 projects using amphibians were on record with the Science Information Exchange in January, 1969. It was these difficulties that led me to maintain laboratory stocks of amphibians, particularly *Rana pipiens*, although it is impossible to be limited to this species. These efforts culminated in the organization of the Amphibian Facility at The University of Michigan, the character and operation of which are described elsewhere.[3]

THE SIGNIFICANCE OF AMPHIBIANS

IN EDUCATION

The numbers of amphibians required for educational use has markedly increased in recent years. This trend is a consequence of rewritten textbooks that call for greater use of living animals in the classroom, coupled with the increased numbers of students. Amphibians occupy a unique position among experimental animals. On the one hand, the rat and mouse, which have long been symbionts of man, have never been revered and are not popular in the high school classroom; on the other hand, as the AVMA well knows, domesticated pets are revered so highly that their use in biomedical research is repulsive to large segments of the population. These factors cannot be ignored in considering the significance of developing amphibian resources for educational use and biomedical research.

However, a serious problem plagues the primary collectors who supply amphibians for educational as well as research purposes. They are mainly centered in Wisconsin and Vermont, with fewer collectors in other parts of the country. It is from this limited number of primary sources that most of the suppliers, who are mainly jobbers, receive their animals.

In nearly all cases, the primary collectors must obtain their animals in violation of state laws that do not permit collection during the seasons when amphibians are in greatest demand for educational and research purposes. Indeed, these laws have not been designed with sufficient understanding of the measures necessary to conserve the natural populations of amphibians. For example, the bullfrog season usually opens just before spawning time. As a consequence of the many difficulties that now affect this industry, young people have not been drawn to it as a field of endeavor, while those who are now supplying the animals are rapidly reaching an age when they will no longer be able to continue their services. An article recently published in *The New York Times*[4] clearly documents the question at issue.

Unless drastic measures are taken promptly to improve both the natural supply of amphibians and the techniques for their collection and distribution, the more than 10,000,000 leopard frogs that are currently supplied each year to the educational and research fields soon will no longer be available. The problem with regard to bullfrogs is even more critical. Without corrective action, the frog industry almost certainly will disappear within the next ten years in spite of growing demands for these animals.

Again, I believe that the AVMA can help by attracting new blood into this small, but vital, education and research industry, and can also help by promoting improved legislation for the conservation of these animals—legislation that will permit the dealers supplying our needs to conduct their businesses within the framework of law. The Amphibian Facility, or other laboratories like it, will not meet the need for animals. The Amphibian Facility was designed to provide defined, laboratory-raised amphibians, not to meet the great demand for animals used in the classroom or for experiments that do not require special material.

IN THE LABORATORY

The extent to which amphibians are used in the research laboratory is only vaguely appreciated because after they are received, most die, some are used, and soon little evidence remains of their presence. Occasionally a cry for help goes out to the veterinarian or zoologist, but to little or no avail. Examination of the Notices of Research Projects recorded in the Science Information Exchange reflects the real character of amphibian use in laboratories. As of January, 1969, 277 such projects were on file, a figure that underestimates the real significance of these animals because they are frequently used on low-budget projects not funded by the agencies that file with the Science Information Exchange.

My analysis of these projects is tabulated in Tables 1–5. Table 1 summarizes the support by animal, character of work, and agency. It does not show that among anurans the leopard frog (*R. pipiens*), the bullfrog (*R. catesbeiana*), and the marine toad (*Bufo marinus*) account

TABLE 1 Support of Research Using Amphibians (Compiled from Science Information Exchange Notices of Research Projects active as of January, 1969)

| | Research Object and Character[a] | | | | | | Total[b] Projects |
| | Anuran | | | Urodele | | | |
Agency	Medical	Basic to Medicine	Nonmedical	Medical	Basic to Medicine	Nonmedical	
FEDERAL AGENCY							
NIH Extramural	62	47	—	14	10	—	125
NIH Intramural	12	—	—	5	—	—	12
PHS Contracts	3	—	—	2	—	—	4
Vet. Admin.	10	—	—	1	—	—	11
NSF	5	19	9	2	18	8	50
AEC	3	6	2	—	7	1	18
U.S. Air Force	2	2	—	2	—	—	5
Smithsonian	2	1	2	2	—	1	5
Dept. Agri.	2	—	—	—	—	—	2
Dept. Inter.	—	1	2	—	—	3	4
NASA	1	—	—	—	—	—	1
Subtotal	102	76	15	28	35	13	237

STATE	2	—	3	—	—	4	6
FOUNDATIONS							
Amer. Heart Assoc.	15	3	—	1	—	—	19
Amer. Cancer Soc.	4	—	—	1	—	—	5
Damon Runyon	2	1	—	—	—	—	3
Amer. Med. Assoc.	2	—	—	—	—	—	2
Musc. Dyst. Assoc.	1	—	—	—	—	—	1
Multiple Scler.	1	—	—	—	—	—	1
Nat. C. Comb. Blind.	1	—	—	1	—	—	2
Amer. Phil. Soc.	—	—	—	—	—	1	1
Subtotal	26	4	0	3	0	1	34
Percent of Projects (34)	76	12	0	9	0	3	—
Total	130	80	18	31	35	18	277
Percent of Projects (277)	47	29	6	11	13	6	—
Object Total	—	228	—	—	84	—	—
Percent of Projects (277)	—	82	—	—	30	—	—

[a]No projects were listed under more than one character.
[b]35 projects used both anurans and urodeles.
[c]Percent of 277.

TABLE 2 Topics of Medically Oriented Research Using Anurans (Compiled from Science Information Exchange Notices of Research Projects active as of January, 1969)

Topic	NIH Extramural	NIH Intramural	PHS Contracts	V.A.	NSF	AEC	U.S.A.F.	Smithsonian	Dept. Agriculture	NASA	States	American Heart Assoc.	American Cancer Soc.	Damon Runyon	American Medical Assoc.	Muscular Dystrophy	Multiple Sclerosis	Nat. Council Combat Blind.	Total[a]
Comp. Neuro. Anat.	3	4	—	—	—	—	—	—	—	—	—	—	—	—	—	—	—	—	7
Physiol. Olfact. Sys.	2	—	—	—	—	1	—	—	—	—	—	—	—	—	—	—	—	—	3
Physiol. Acoust. Sys.	1	—	—	—	—	—	1	—	—	—	—	—	—	—	—	—	—	—	2
Physiol. Vision	4	1	1	1	1	—	1	—	—	—	—	—	—	—	—	—	—	—	9
Spin. Cord Test Sys.	—	—	—	—	—	—	—	—	—	—	—	—	—	—	—	—	1	—	1
Electrophysiol.—																			
Non-sens. Organs	7	—	—	—	—	2	—	—	—	—	—	1	—	—	—	—	—	—	10
Conditioning	—	—	—	1	—	—	—	—	—	—	—	—	—	—	—	—	—	—	1
Psychoactive Drugs	1	—	—	—	—	—	—	—	—	—	—	—	—	—	—	—	—	—	1
Ion Trans. & Regul.	18	4	—	4	1	—	—	—	1	—	1	5	—	—	—	—	—	1	35
Kidney Physiol. &																			
Pharm.	2	—	—	—	—	—	—	—	—	—	—	1	—	—	—	—	—	—	3
Biochem. & Pharm.—																			
Skel. & Heart Mus.	9	1	—	—	—	—	—	—	—	—	—	6	—	—	—	1	—	—	17

Topic																			Total
Biochem. & Pharm.—Smooth Muscle	2	—	—	—	—	—	—	—	—	—	—	—	—	—	—	—	—	—	2
Environment on Organ Physiol.	—	—	—	—	—	—	—	—	—	—	—	—	1	—	—	—	—	—	1
Endocrinology	5	—	—	1	—	—	—	—	—	—	—	—	—	1	—	—	—	—	7
Biochem. of Hemoglob.	1	—	—	—	—	—	—	—	—	—	—	—	—	—	—	—	—	—	1
Blood Clotting	1	—	—	—	—	—	—	—	—	—	—	—	1	—	—	—	—	—	2
Lungs	1	—	—	2	—	—	—	—	—	—	—	—	—	—	—	—	—	—	3
Nicotine Action	—	—	—	—	—	—	—	—	—	—	—	—	—	—	2	—	—	—	2
Pesticide Toxicity and Pharm.	2	—	1	—	—	—	—	—	—	—	—	1	—	—	—	—	—	—	4
Amphibian Toxins	—	1	—	—	—	—	—	—	—	—	—	—	—	—	—	—	—	—	1
Parasitology	2	—	—	—	—	—	—	—	—	—	—	—	—	—	—	—	—	—	2
Virology	1	1	—	—	—	—	—	—	—	1	—	—	—	—	—	—	—	—	3
Immunol. & Tissue Transplantation	2	—	—	1	3	—	—	—	—	—	—	—	1	—	—	—	—	—	7
Cancer	5	—	1	—	1	—	—	1	—	—	—	—	2	2	—	—	—	—	12
Wound Healing	2	—	—	—	—	—	—	—	—	—	—	—	—	—	—	—	—	—	2
Physiol. of Develop.	—	—	—	—	1	—	—	—	—	—	—	—	—	—	—	—	—	—	1
Teratology	—	—	—	—	—	—	—	1	—	—	—	—	—	—	—	—	—	—	1
Magnetic Field Action	—	—	—	—	—	—	—	—	—	1	—	—	—	—	—	—	—	—	1
Totals[a]	71	12	3	10	7	3	2	2	2	1	2	15	4	2	2	1	1	1	141

[a]Total not equal to Table 1 because of "Group Applications" covering several projects. Each project is recorded under one topic only.

TABLE 3   Topics of Medically Oriented Research Using Urodeles (Compiled from Science Information Exchange Notices of Research Projects active as of January, 1969)

Topic	NIH Extramural	NIH Intramural	PHS Contracts	V.A.	NSF	U.S.A.F.	Smithsonian	American Heart Assoc.	American Cancer Soc.	Nat. Council Combat Blind.	Total[a]
Comp. Neuro. Anat.	1	5	—	—	—	—	—	—	—	—	6
Physiol. Olfact. Sys.	1	—	—	—	—	—	—	—	—	—	1
Physiol. Vision	—	—	—	—	—	1	—	—	—	1	2
Kidney Physiology and Pharm.	1	—	—	—	1	—	—	—	—	—	2
Biochem. & Pharm.— Skel. & Heart Mus.	—	—	—	—	1	—	—	—	—	—	1
Genetics, Heart Beat	—	—	—	—	—	—	—	1	—	—	1
Endocrinology	3	—	—	—	—	—	—	—	—	—	3
Radiosensitivity	1	—	—	—	—	—	—	—	—	—	1
RBC Pathology	1	—	—	—	—	—	—	—	—	—	1
Nicotine Action	—	—	1	—	—	—	—	—	—	—	1
Parasitology	1	—	—	—	—	—	—	—	—	—	1
Virology	1	—	—	—	—	—	—	—	—	—	1
Immunol. & Tissue Transplantation	3	—	—	—	—	—	—	—	—	—	3
Cancer	—	—	1	1	—	—	1	—	—	—	3
Regeneration	2	—	—	—	—	—	—	—	1	—	3
Teratology	—	—	—	—	—	—	1	—	—	—	1
Laser Effects	—	—	—	—	—	1	—	—	—	—	1
Totals[a]	15	5	2	1	2	2	2	1	1	1	32

[a]Total not equal to Table 1 because of "Group Applications" covering several projects. Each project is recorded under one topic only.

for 90 percent of the animals used, or that among urodeles, the axolotl (*Ambystoma mexicanum*) accounts for 30 percent of the effort. Tables 2–4 identify the research topics under investigation and Table 5 shows the location of the work.

Table 1 shows that 228 of these projects utilized anurans, and 84 utilized urodeles (35 used both): 58 percent were directly related to medicine, 42 percent were devoted to studies basic to medicine, and 12 percent were not related to medicine (not equal to 100 percent because the columns under anurans and urodeles are added and some projects used both). Information on the dollar value of this investment was not available. The significance of amphibians as tools in biomedical research can scarcely be questioned, however.

Several points of interest emerge from an examination of these tables. In spite of the definite orientation of this work toward medicine and medically-related problems, only the work of the Amphibian Facility of The University of Michigan has been directed toward improving the quality of the animals, although others have now become active. The topics under investigation broadly cover the spectrum of medical concern; of special interest, however, is the number of projects concerned with ion transport and regulation. In Table 4 are found the topics that forecast the future of medical research; this table reveals both strengths and weaknesses in our current research effort. The number of projects devoted to endocrinology, to development, and to biochemistry of the nucleic acids, enzymes, and proteins reflects the utility of amphibians, and reveals areas of major interest that require much additional study before findings can be applied clinically.

To understand why amphibians are popular for endocrinological studies, one need only recall that for metamorphosis of the anuran from the tadpole to juvenile stage extensive morphological and physiological changes occur during a relatively short period. Since all the changes associated with metamorphosis are ultimately dependent upon hormonal regulation, this system is ideal for analyzing the mechanisms of endocrine action. Developmental and biochemical studies of various kinds are carried out on amphibian materials because all stages of development occur external to the body and are readily accessible; the materials can, therefore, be obtained in sufficient quantity to permit the necessary biochemical manipulations.

One weakness in the national research effort, as revealed by this table, is in genetics. Few studies on amphibians have been directed toward analysis of their genetics; in fact, one of the reasons for establishing the Amphibian Facility was to help fill this gap. Recall that, in addition to the ease of accessibility of all developmental stages from fertilization on, the genetic composition of amphibians can be controlled by several strategems. It is possible, for example, to obtain haploid animals by parthenogenesis and to examine their development under the influence of a diminished genome. The diploid state of haploid eggs can be reconstituted by using the genome of the second polar body[3] or by suppressing cytokinesis at the first cleavage division. The former permits immediate determination of cross-over frequencies[5]; the latter produces totally homozygous animals. Similarly, tetraploids

TABLE 4 Topics of Research Basic to Medicine Using Anurans and Urodeles
(Compiled from Science Information Exchange Notices of Research Projects active
as of January, 1969)

| | Anurans | | | | | | | | | Urodeles | | | |
| | Agency | | | | | | | | | Agency | | | |
Topic	NIH Extramural	NSF	AEC	U.S.A.F.	Smithsonian	Dept. Interior	American Heart Assoc.	Damon Runyon	Total[a]	NIH Extramural	NSF	AEC	Total
Comp. Neuro. Anat.	2	—	—	1	—	—	—	—	3	—	—	—	—
Physiol. Acoust. Sys.	3	—	—	—	—	—	—	—	3	—	—	—	—
Physiol. Vision	3	1	—	1	—	—	—	—	5	—	—	—	—
Electrophysiol.—													
Non-sens. Organs	5	—	—	—	—	—	—	—	5	1	—	—	1
Orientation & Behavior	1	2	—	—	1	—	—	—	4	—	2	—	2
Ion Trans. & Regul.	3	1	1	—	—	—	—	—	5	—	—	—	—
Nitrogen Metabolism	—	3	—	—	—	—	—	—	3	—	2	—	2
Biochem. & Physiol.—													
Skel. & Heart Mus.	6	—	1	—	—	—	2	—	9	1	2	—	4

Biochem. of Pigments	—	—	—	—	—	—	—	—	—	—	1	1	1
Endocrinology	10	1	—	—	—	—	—	—	11	1	3	—	6
Biochem. of Hemoglob.	1	—	—	—	—	—	—	—	1	—	—	2	—
Fertil. & Reproduction	4	2	1	—	—	—	—	—	7	—	1	—	1
Lungs	—	1	—	—	—	—	—	—	1	—	—	—	—
Biochem. of DNA, RNA, Enzymes & Proteins	9	3	3	—	—	—	—	1	16	2	1	—	3
Genetics	6	1	—	—	—	—	—	—	—	1	1	—	3
Amphibian Toxins	—	—	—	—	—	—	—	—	—	1	—	1	1
Parasitology	1	2	—	—	—	—	—	—	3	—	1	—	1
Virology	—	—	—	—	—	1	—	—	1	—	—	—	—
Immunol. & Tissue Transplantation	1	1	1	—	—	—	—	—	3	—	—	1	1
Cell Division	—	1	—	—	—	—	—	—	1	1	1	1	3
Regeneration & Wound Healing	2	—	—	—	—	—	—	—	2	2	1	2	5
Development	16	4	4	—	—	—	1	1	26	5	3	1	9
Amphibian Facility	—	1	—	—	—	—	—	—	1	—	1	—	1
Totals[a]	73	24	11	2	1	1	3	2	117	15	20	9	44

[a]Totals not equal to Table 1 because of classification of projects under more than one topic, e.g., *Biochemistry of DNA-RNA* and *Development*.

113

TABLE 5 Location of Research Using Amphibians (Compiled from Science Information Exchange Notices of Research Projects active as of January, 1969)

Agency	Clinical	Basic	School Pub. Health	Vet. Med.	Zoology & Psychology	Chemistry	Res. Center	Agri. Exp. Station	Museum	Dept. Unspecified	Intramur. or Institute	Hospital	Museum	Total
		Medical			Non-Medical							Non-Academic		
FEDERAL AGENCY														
NIH Extramural	15	44	1	1	45	4	1	—	1	—	6	2	—	120
NIH Intramural	—	—	—	—	—	—	—	—	—	—	12	—	—	12
PHS Contracts	—	—	—	—	—	—	—	—	—	2	2	—	—	4
Vet. Administration	3	2	—	—	—	—	—	—	—	—	—	8	—	13
NSF	1	14	—	—	36	—	—	—	1	—	1	—	1	54
AEC	1	3	—	—	4	—	2	—	—	—	7	—	—	17
U.S.A.F.	—	2	—	—	2	—	—	—	—	—	1	—	—	5

Smithsonian	—	—	—	—	—	—	—	—	—	—	2	—	3	5
Dept. Agriculture	1	—	—	—	—	—	—	—	—	—	1	—	—	2
Dept. Interior	—	—	—	—	1	—	—	—	—	—	3	—	—	4
NASA	—	—	—	—	—	—	—	—	—	—	1	—	—	1
STATE	—	—	—	—	—	—	—	5	—	—	—	—	—	5
FOUNDATIONS														
Amer. Heart Assoc.	2	13	—	—	3	—	—	—	—	—	1	—	—	19
Amer. Cancer Soc.	—	3	—	—	2	—	—	—	—	—	—	—	—	5
Damon Runyon	—	—	—	—	3	—	—	—	—	—	—	—	—	3
Amer. Medical Assoc.	—	2	—	—	—	—	—	—	—	—	—	—	—	2
Musc. Dyst. Assoc.	1	—	—	—	—	—	—	—	—	—	—	—	—	1
Multiple Sclerosis	1	—	—	—	—	—	—	—	—	—	1	—	—	2
Nat. C. Comb. Blind.	1	—	—	—	—	—	—	—	—	—	1	—	—	2
Amer. Phil. Soc.	—	—	—	—	1	—	—	—	—	—	—	—	—	1
Total	26	83	1	1	97	4	3	5	2	2	39	10	4	277
Percent of Projects (277)	9	30	0	0	35	1	1	2	1	1	14	4	1	
	39%				41%						19%			

or other chromosomal combinations can readily be engineered. In addition to these manipulations of the genome, it is possible to transfer nuclei from embryos at advanced stages of development to enucleated eggs and to obtain normal progeny. With appropriate experimental designs, clones of identical, heterozygous animals can be produced. No other vertebrate is amenable to such a variety of genetic manipulations. How much more information amphibians might yield if as many of their gene loci were identified as are known in the mouse!

Genetic engineering is advertised as an important biomedical area for the future. These tables show no genetic engineering projects, yet it is only with amphibian material that we have the technology necessary to initiate immediately preliminary tests of this concept. I should like to draw attention to the work of Berns and co-workers,[6, 7] who have produced specific lesions in the chromosomes of amphibian cells in culture by using irradiation by laser beam. It will be exciting to determine what genetic consequences result when nuclei with such controlled chromosomal defects are transferred from the tissue culture cells to enucleated eggs that are then permitted to develop. These tables do not explicitly reveal the excitement and potential of the future.

Table 1 shows another significant fact. Typically, projects of the American Heart Association are strongly oriented toward practical goals. It supports only one project using the urodele. Examination of the abstract reveals that the research animals were axolotls carrying a mutant that affects heart development. Certainly, as amphibian strains are developed and genetic control is established, the amphibian can be expected to play a much greater role in clinically oriented biomedical investigations.

Other areas of weakness are revealed by these tables. Only two projects are directly concerned with teratogenesis—the study of developmental defects. It is surprising that amphibians are not used more in this area, not only because the embryo is ideally suited to external manipulation, but also because we know a great deal about the chemistry of its development.[8] Only eight projects are concerned with regeneration and wound healing, another major area of weakness. It is a common dictum of biology that the animal chosen for research should be the one ideally suited to the question under examination. Among the amphibians, the urodeles show an extensive capacity for regeneration, whereas the anurans do not. Certainly it is by comparing these animals with each other, and with higher vertebrates, that we might discover the secrets of replacing lost limbs or organs by regeneration.

Considering the large research and clinical investment in organ replacement, I find it most surprising that a greater number of our most imaginative minds do not address themselves to the question, however difficult, of organ replacement by regeneration, rather than by transplantation.

Amphibians are valuable in many other areas of biomedical research. Because there are great differences in their physiology between summer and winter, many specific physiological events can be investigated by conducting experiments designed to test the influence of temperature. This variable generally cannot be investigated using homeotherms. Finally, the position of amphibians at the evolutionary transition point from aquatic to terrestrial modes of living suggests many areas for comparative study that could lead to improved understanding of the biology of all animals.

AMPHIBIAN FACILITY RECORDS

No program of animal study can be meaningful unless serious attention has been given to records. It is the records that control input, output, and priorities of daily effort. If records are poor, the entire program can be lost; if complete and efficient, the return on the investment can be multiplied manyfold. Because the Amphibian Facility represents the first attempt to develop defined laboratory strains of amphibians, we have the responsibility, as well as the rare opportunity, to initiate record systems that will yield maximal returns from the energy expended and, we hope, establish useful traditions for other amphibian colonies.

The information we record falls into three general categories: (1) demographic, (2) genetic, and (3) pathologic. Currently the system is designed around Termatrex equipment. In this system[9, 10] each item of information is represented by a plastic card in which 10,000 holes can be drilled. Each hole represents the serial number of a particular animal, and is identified by its grid position on the card.

Table 6 shows the sheet on which the demographic information is recorded. Each batch of animals is assigned a serial number, and the pertinent information is recorded on a sheet of this kind. When an animal from that batch takes on individual significance, it receives an individual serial number, and the information describing that animal is recorded. The random-access plastic cards are then drawn from the file. Thus, for an animal from nature, the group of cards might be

TABLE 6 Individual Frog Record

Species_____________________________________ Serial (T)#_____________________W30

    B00 01 02 03 04 05 06 07 08 09 10 11 12 13 14 15 16 17 18 19

Cage #_____________________

W80   University Owned

Origin

W27   Identification uncertain

W00 *Not Amphibian Facility*
W01*   Commercial__________
W02*   Private__________
(*See P00–12 for breakdown)
B20–39 Geographical__________
B40–53
65–74   Date acq.__________

    Purchase Order #__________
    Length acq.__________ mm
    Weight acq.__________ gm

W03 *Amphibian Facility*
B40–53
65–74   Date fert.__________
G00–39
80, 82 Mother T#__________
G40–79
81, 83 Father T#__________
W04   Biparental
W05   Hybrid
W28   "Sperm irradiated"
W06   Gynogenetic
W29   "Homozygous diploid"
W07   Other__________
S40–53
65–74   Metamorph. Date__________
P56   Length at Metam.__________ mm
P57   see Growth Card

Use

As Parent (♀—see Mult. Ov. Cd.)
          (♂—use Growth Cd.)

W81   1°, Date_____ Prog. #_____
W82   2°, Date_____ Prog. #_____
W83   3°, Date_____ Prog. #_____
W84   4°, Date_____ Prog. #_____
W85   5°, Date_____ Prog. #_____
W86   6°, Date_____ Prog. #_____
W87   More than 6 attempts

Health (see Path Record)

Identification

W08   Male   W09   Female
W10   Toe Clipped   Picture
      FL__________ FR__________ P54
      HL__________ HR__________
W11   Wild Type
W14   Dark
(For specifics—see Genotype Record)
W17   Other color__________

W22   Other pattern__________

P55   Color picture
P58   X-ray picture

W23   Lucke   B78 Lucke x –   B79 L x L

(For abnormalities—see Path Record)

Color Pattern

W31–39 Background__________
W40–49 Head__________
W50–59 Body__________
W60–69 Legs__________
W70–79 Other__________

Special Use

       Grafting (see Grafting Card)
B75   As Host   B76   As Donor
B77   Test procedures used__________

B54   Electrophoresis
B55   Used as gynogenetic mother
B56   Testis removed or ovaries examined
B57   Tumor fraction injected into embryo
B58   Tumor fraction treated with lysozyme
B59   Tumor fraction into larva
B60   Biopsy (also see Path Record)
B61–64 Other tests to be listed

Disposal

W94   Disposed (Punch red transparent cd.)
      (Death—see Path Record)
W97   Donated: To:__________
W98   Sold: To:__________
      Invoice #__________

"white card 00" (W 00—not Amphibian Facility), "W 02 (from a private source), appropriate black cards to indicate the geographic origin and date of acquisition, "W 08" to indicate male, and "W 11" to describe a wild-type animal. These cards are then simultaneously punched with a precision drill at the grid position corresponding to the serial number of the animal. If any card is placed on a luminescent box, all punched grid positions become visible and can be counted to give an inventory of all animals in the colony that are described by that characteristic. If additional cards are added, questions can be answered. Thus, for example, using such descriptors, one could ask: How many animals raised in the Amphibian Facility (W 03), from a fertilization in 1967 (B 67), with the Burnsi color pattern (W 18), were used as a parent at least once (W 81)? The answer to this question happens to be 37.

Similarly, the genetic background of the animal is recorded on the basis of the descriptors found in Table 7, and, if an animal becomes sick and is treated or dies, the pertinent information is recorded using the descriptors shown in Table 8. Examination of these tables, however, will reveal a great deal about the operation of the Amphibian Facility and about the information we are collecting. The master code that lists all descriptors used in the system is not reproduced, but may be obtained from the author.

Of particular help in this effort has been the assistance rendered the Amphibian Facility through the cooperation of the Unit for Laboratory Animal Medicine of The University of Michigan, of which Dr. Bennett J. Cohen is director. Representatives of this unit have worked closely with the Amphibian Facility in identifying pathological conditions affecting our animals and in suggesting husbandry techniques to alleviate these ills.[11] This laboratory's work finds concrete expression in the pathological data recorded according to Table 8.

Use of this system has permitted us to control the development of the Amphibian Facility and to accumulate extensive data that has been or will be the subject of publications on the normal and abnormal biology of amphibians.[5,12,13] It has also enabled us to identify information items that are useful in practice, and to discard items with only slight retrieval value.

There are limitations on the Termatrex system. One deck of cards is good for only 10,000 animals—a capacity that was sufficient for the first several years of development of the Amphibian Facility. Only a few months of actual operation were needed, however, to assign all numbers for animals from 10,000 to 20,000, and already numbers above

TABLE 7   Individual Genotype Record

Serial (T) #________________

Genotype				*Phenotype*	
W18	*Burnsi* (Phenotype, tbt)			W12	*Green*
	Prob. of homozygosity______%			S12	Green in pedigree, tbt
bl(T)	Tested			bl(T)	Tested
S00	B/B				
S01	B+/B			W13	*Brown*
S02	B+/B+			S14	Brown in pedigree, tbt
				bl(T)	Tested
W19	*Kandiyohi* (Phenotype, tbt)				
	Prob. of homozygosity______%			W15	*Blue*
bl(T)	Tested			S16	Blue in pedigree, tbt
S03	K/K			bl(T)	Tested
S04	K+/K				
S05	K+/K+			W20	*Tessellated*
				S18	Tessellated in pedigree, tbt
	Albino			bl(T)	Tested
S06	In pedigree, tbt				
	Prob. of heterozygosity______%			W23	*Lucké tumor*
bl(T)	Tested			S20	Lucké tumor in pedigree, tbt
S07	a+/a+			bl(T)	Tested
S08	a+/a				
W16	a/a			S22	*Double albumin*
				S23	Double albumin in pedigree, tbt
	Melanoid			bl(T)	Tested
S09	In pedigree, tbt				
	Prob. of heterozygosity______%			S25	*Hb-less*
bl(T)	Tested			S29	Hb-less in pedigree, tbt
S10	m+/m+			bl(T)	Tested
S11	m+/m				
W21	m/m			S28	*Spinner*
				S29	Spinner in pedigree, tbt
	"Alleles"			bl(T)	Tested
S76	m$_1$	S80	m$_5$		
S77	m$_2$	S81	m$_6$	S75	Potential Marker ______________
S78	m$_3$	S82	m$_7$		(Record in log)
S79	m$_4$	S83	m$_8$		
S36	*Dark* (Phenotype, tbt) (axolotl)				
	Prob. of homozygosity______%				
bl(T)	Tested				
S37	D/D				
S38	D/d				
S39	d/d, white				

Note:	bl(T)	=	Transparent blue card labeled with name of marker.
			To be punched when genotype is tested.
	tbt	=	To be tested.

TABLE 8 Individual Amphibian Pathology Record

Species_________________________________ (B00-19) Serial (T) #_________________________

W93 Necropsy Report #_______________________________ Report Date_________________________

By____

| W00 | Wild | W03 | Lab raised | | W08 | Male | W09 | Female |

Character of Necropsy
B80 Minimal
B81 Routine
W92 Complete

W88 Illness, Date 1st noted:____________
P40–53
  65–74 Date of death ___________________
Postmortem interval _______________________

Age (see B or S 40–53, 65–74)
Bu00 Unknown
Bu01 Premetamor. (Stage) _______________
Bu02 Metamor. (Stage) _________________
Bu03 <1 mo. Postmeta. (mo.)___________
Bu04 1–< 6 mo. Postmeta. (mo.)________
Bu05 6–<12 mo. Postm. (mo.)___________
Bu06 1–<2 yr. Postm. (yr.)_____________
Bu07 2–<3 yr. Postm. (yr.)_____________
Bu08 3 or more yr. Postm. (yr.)_________
    (For wild animals use time in A.F.)

Mode of Death
W95 Killed
W96 Spontaneous

Length at death________________(HR/HT) mm
Weight at death ____________________________ gm

B88 Picture of disease
P58 X-ray of animal
B60 Biopsy

Animal Location when Diseased
Bu09 Hibernation   Bu10 Room temp.
Bu11 In group _____________________________
Bu12 In isolation ________________________

Correlation of Disease
Bu13 No significant event known
Bu14 Reproduction associated
Bu15 In anesthesia—post operative
Bu16 Experimental animal_______________
Bu17 Injury
Bu18 Other_____________________________

CLINICAL EXAMINATION

Appearance and Behavior

Bu23 Normal
Bu24 Not eating
Bu25 Depressed—lethargic
Bu26 Abnormal posture
Bu27 Abnormal swimming
Bu28 Abnormal locomotion

Duration of Illness (circle choice)
Bu19 Unknown
Bu20 0–24 hrs. or unknown 0–48 hrs.
Bu20+
21 24–48 hrs.
Bu21 48–72 hrs. or unknown 24–96 hrs.
Bu21+
22 3–4 days
Bu22 >4 days or unknown >3 days

Abnormal Fluid Accumulation
Bu29 Local swelling
Bu30 Generalized swelling (edema)
Bu31 Blood in coelomic cavity
    (seen through abdominal wall)

Abnormal Skin
Bu32 Abnormal shedding
Bu33 Skin discoloration:________________
Bu34 Lesion(s):_________________________
Bu37 Abdominal enlargement
Bu38 Bleeding from body opening
Bu39 Prolapse of
Bu40 Palpable mass
Bu41 Dehydration

Abnormality

	Not Congenital		Congenital		Of
Bu46	_________________	Bu47	_________________		Ventral Surface
Bu48	_________________	Bu49	_________________		Dorsal Surface
Bu50	_________________	Bu51	_________________		Eye

TABLE 8 (Continued)

Abnormality (continued)

Not Congenital		Congenital		Of
Bu52	_______	Bu53	_______	Ear-Nose
Bu54	_______	Bu55	_______	Mouth
Bu56	_______	W24	_______	Forelimb
Bu57	_______	W25	_______	Hindlimb
Bu58	_______	Bu59	_______	Feet
Bu60	_______	Bu61	_______	Spine
Bu62	_______	W26	_______	Other

W89   *Provisional Diagnosis* from clinical exam. (Describe unusual circumstances associated with the disease)____

____

____

____

____

W90   *Treatment:*
Bu65 None   Bu66 Change of Environment: (Describe)___________

Bu67 Isolated   ____

Bu68 Chloromycetin   Bu69 Terramycin

Bu70 Other antibiotic: (Identify, describe)____

____

Bu72 Other drugs: (Identify, describe) ____

____

Bu74 Surgical: (Identify, describe)____

____

Result of Treatment:
Bu76 Ineffective: (Describe) ____

____

Bu77 Effective: (Describe) ____

____

DIAGNOSTIC PROCEDURES (Applies to clinical exam also; see supplemental reports for details)

B82	Biochemistry	B85	Histopathology	B87	Photo, gross of disease
B83	Gross Pathology	W91	Microbiology	B88	Photo, microscopic
B84	Hematology	B86	Parasitology	P58	Photo, x-ray

NECROPSY EXAMINATION

B89   Lesion(s) of skin or other organ(s): (Identify) _______________________________
    Elevated   Depressed   Color (red, white, yellow, black,____)
    Size of lesion ($<$1mm; 1–2mm; $>$2mm $<$1.0cm; 1–2cm; $>$2cm)
    Describe:____

B90   *Hemorrhage:* (pinpoint, 0.5cm; 1.0cm; $>$1.0cm; extensive; blood in coelom)
    Describe: ____

B91   *Swelling:* (localized; organ enlargement; generalized swelling)
    Describe: ____

B92   *Excessive fluid accumulation in coelom*

B93   *Excessive pigmentation:* (Describe)____

B94   *Neoplasia:* (Use W23 for Lucke and complete the following)
    Location:_______________ Possible metastasis:______________________
    Size:________gm;________mm Color:___________Consistency:____________

Site of pathology (S=system; identify structure)
Bu81 Skin and subcutaneous ___________________________
Bu82 Musculoskeletal S.___________________________
Bu83 Respiratory S. ___________________________

TABLE 8 (Continued)

Bu84 Cardiovascular S.___________________________
Bu85 Hemic & lymphatic S.___________________________
Bu86 Digestive S.___________________________
Bu87 Urogenital S.___________________________
Bu88 Endocrine S.___________________________
Bu89 Nervous S.___________________________
Bu90 Special sense organ ___________________________
Bu91 Non-specific___________________________
Bu92 *Viral-bacterial-fungal*   Bu93 *Protozoan parasite*   Bu94   *Metazoan parasite*
  In:
Bu95 Head: mouth; eye; ear; nose; eustachian tube; gill; other:
Bu96 Thorax: esophagus; heart; lung; coelom; other:
Bu97 Abdomen: stomach; small-large intestine; liver; gall bladder; kidney; urinary bladder;
        gonad; reproductive ducts; coelom; other:
Bu98 Other: skin; connective tissue; blood vessel; lymph spaces; other:

B95 *Tissues saved at necropsy*
W99 Whole animal preserved: (Code following only if preserved separately)
B96 Head: oral structures; eye; ear; nose; eustachian tube; gill; brain;
       other:
B97 Thorax: esophagus; heart; lung; other:
B98 Abdomen: stomach; small-large intestine; feces; liver; gall bladder; spleen; kidney;
       urinary bladder; urine; fat body; gonad; reproductive ducts; other:
B99 Other: skin; skeletal muscle; bone; artery; vein; other:

Bu99 *Diagnosis* from full data: (Use pathology terms from list in code book. Include
        "evaluation." Use back of sheet if necessary.)

20,000 are being allotted. At this stage we must begin the transition to automatic storage and retrieval of information by computer. Several systems that are under consideration will allow more rapid statistical analyses of the data and will greatly simplify surveillance of the accuracy of the information being collected.

At this early stage of development of the Amphibian Facility, we welcome suggestions from those of you who have had experience with the utilization of information generated by animal facilities. We hope your suggestions will aid us in improving the services we can perform, not only for those working within the Amphibian Facility, but also for those in the general scientific community who may wish to draw upon our information. When we have completed the transition to a computer system, it would be useful to incorporate data collected by any scientist on the normal and abnormal biology of amphibians of any species and to make this information available to all who request it. I would urge that those of you who do have experience with amphibians, or who will have experience in the future, communicate with us concerning the possibility of incorporating portions of your data into this information bank. Hopefully, this practice will lead to earlier solutions to many of the remaining problems of amphibian husbandry.

This work was supported by National Science Foundation grants GB 4677X and GB 8187X, by U.S.–Japan Cooperative Science Program, National Science Foundation grant GF 242, by Damon Runyon Memorial Fund grant DRG 980, by The University of Michigan grant ORA 36608, by Michigan Cancer Research Pilot Projects #4 and #17, by National Science Foundation Institutional grant #108, and by National Institutes of Health grant 5 T01 GM0989.

REFERENCES

1. Wightman, W. P. D., 1950. The growth of scientific ideas. p. 220. Oliver and Boyd, Edinburgh.
2. Nace, W. G. 1970. Specific macromolecular changes during development. *In* Cell Differentiation. O. A. Schjeide and J. de Vellis, eds. Chapter 12. D. Van Nostrand, Reinhold Co., New York. In Press.
3. Nace, G. W. 1968. The Amphibian Facility of the University of Michigan. BioScience 18(8):767–775.
4. Sterba, J. P. 1969. "Lo, the poor bullfrog! A victim of progress." The New York Times, April 23, 1969, p. 49.
5. Nace, G. W. and C. M. Richards, 1969. Development of biologically defined strains of amphibians. *In* Biology of Amphibian Tumors. M. Mizell, ed. Springer-Verlag, New York. pp. 409–418.
6. Berns, M. W., R. S. Olson, and D. E. Rounds, 1969. *In vitro* production of chromosomal lesions with an argon laser microbeam. Nature 221:74–75.
7. Berns, M. W. and D. E. Rounds. 1970. Cell surgery by laser. Sci. Amer. 222(2):98–110.
8. Nace, G. W. Immunochemical analysis of development in the frog: Illustration of a model system for studies of teratogenesis. *In* 2° and 3° Berkeley and Boulder Workshops. W. Asling, M. Runner, and J. Wilson, eds. In Press.
9. Carpenter, H. M., 1962. A system for storage and retrieval of data from autopsies. Am. J. Clin. Pathol., 38(5):449–467.
10. Systems Manual, 1967. Jonker Corporation, Gaithersburg, Maryland.
11. Cohen, B. J. 1969. Laboratory animal medicine in relation to transplantation biology. Tijdschr. Diergeneesk. 94(1):28–32.
12. Joiner, G. N. and G. D. Abrams, 1967. Experimental tuberculosis in the leopard frog. J. Am. Vet. Med. Assoc., 151(7):942–949.
13. Abrams, G. D., 1969. Diseases in an amphibian colony. *In* Biology of Amphibian Tumors. M. Mizell, ed. Springer-Verlag, New York. pp. 419–428.

Animal Models for
Pharmacotherapeutic Studies

DAVID P. RALL

Certain difficulties arise when one attempts to apply knowledge gained in the study of a particular disease, and its response to therapy (drug therapy in this instance), in one species to an identical or similar disease in a different species. I shall attempt to consider these problems, first, as they relate to the drug and, second, as they relate to one group of diseases, cancer. The general principles involved, I am convinced, will apply to other groups of diseases with which I am much less familiar.

PROBLEMS RELATED TO THE DRUGS

This first topic might be subtitled, "Extrapolation of the Results of Studies on Drug Action from Laboratory Animals to Man." This is a difficult subject to consider in a scientific, rigorous fashion and, therefore, has generally been treated in an anecdotal manner. Any pharmacologist or toxicologist worth his salt can, and frequently does, remember, discuss, and clearly document individual instances of either a perfect or perfectly imperfect correlation between results in laboratory animals and man. Whenever interspecies comparisons are considered, someone notes that penicillin is so toxic in the guinea pig that, had it first been tested in this species, it never would have been developed further.[2, 13] If I too may be anecdotal, let me discuss this case; the facts are interesting as always. First, it is clear that while the single dose toxicity in guinea pigs, for penicillin, is about the same as in other animals,[7] the apparent differences or discrepancies

125

occur after repeated doses in subacute toxicity studies. At relatively
low repeated doses, significant mortality occurs in guinea pigs. This
is generally in the order of 50–80 percent and rarely, if ever, 100 percent. Further, the guinea pigs become tolerant to this toxic effect and
newborn or very young guinea pigs do not show this unusual susceptibility. A careful examination of the mechanism of this mortality has
been under way for a number of years and it seems to involve in some
way superinfection and/or bacterial overgrowth in the guinea pig.[3]
This effect of penicillin on the bacterial flora of the GI tract of the
guinea pig is secondary. Leptospirosis in guinea pigs ordinarily is lethal,
but this disease can be cured by using penicillin. Thus, toxicity is not
so prohibitive that penicillin is therapeutically ineffective.[8] More important, the phenomenon of bacterial superinfection is seen in man,
and is one of the major clinical problems associated with the use of
antibiotics. Thus, the observation of the toxicity of penicillin in the
guinea pig did, in fact, predict a serious toxic effect of penicillin and
other antibiotics in man. It seems to me that the guinea pig "predicted"
very well. I am more concerned about the pharmacologist's failure to
appreciate the prediction.

Before considering extrapolation of biological data from one species
to another, we must define a "species." A species is a group of individuals with a number of common characteristics that are different from
those of other groups of individuals. Variations around these common
characteristics are not in large quantum jumps, but are continuously
variable. There are differences due to sex and age, of course, as well
as those due to specific genetic polymorphisms. Thus, each species,
by definition, has certain unique characteristics. It would be surprising, indeed, if some aspects of this uniqueness were not related to responses to exogenous chemical agents.

A number of steps are involved in the pharmacological response of
an animal to a drug. These steps intervene between drug administration
and the arrival of the drug in adequate concentration at its site of action. They include absorption, distribution, metabolism, excretion
and effect at ultimate site(s) of action. It is important to analyze these
steps in relationship to the degree of consistency among various species.
First, we will consider absorption. Most drugs are given orally; therefore, it is important to consider absorption after oral administration.
In general, laboratory animals and man are very similar in this regard.
Studies by Hogben[11] and Schanker[18] are classic and demonstrate this
phenomenon in detail. Hume and co-workers[12] studied the absorption
of thiopental and pentobarbital from the GI tract of laboratory animals

and man. With thiopental, peak plasma concentrations and time courses are essentially similar between man, dog, and rat. However, for pentobarbital, which is less lipid-soluble, absorption in man occurs later and slower than for the others. Thus, a quantitative species difference seems to exist. Welch has described the rapid absorption of 6-azauridine in laboratory animals compared with man. This compound is completely absorbed after oral administration in mouse, rat, and dog, but only about 25 percent in man. A further interesting aspect is that in man unabsorbed azauridine is cleaved in the large bowel by bacterial action to the free base 6-azauracil, which then can be absorbed. Unfortunately 6-azauracil, unlike azauridine, causes certain central nervous system symptoms that can be observed after oral administration of azauridine in man.[19] In newborn animals of different species, there is a clear difference between their abilities to absorb intact protein directly from the GI tract. Antibodies can be absorbed directly by suckling mice, rats, cattle, goats, sheep, pigs, horses, and dogs, whereas guinea pigs, rabbits, and human beings do not normally absorb whole protein after birth.[6]

With respect to renal excretion there are known differences in the handling of drugs. The most striking of these variations occurs in marine fish. A very primitive cyclostome, *Myxine glutinosa*, the hagfish, as an essentially atubular kidney.[17] Urine plasma concentration ratios for phenol red and for inulin are unity. On the other hand, a teleost, *Lophius*, the goosefish, has no functioning glomeruli; therefore, lipid-insoluble compounds will not be excreted into the urine unless they are actively secreted by the renal tubule.[14] It is well known that urinary pH influences excretion of weak acids and bases by nonionic diffusion. Carnivores generally excrete an acid urine, and herbivores an alkaline urine. Thus, diet can affect urinary pH and, therefore, the rate of excretion of drugs that are partially ionized at body pH. Preliminary studies have also shown that biliary excretion can vary from species to species. With methotrexate, a folic acid antagonist, biliary excretion is a prominent feature in the mouse and rabbit while it is not particularly prominent in the rat or in man.[9, 10]

Distribution and storage of the drugs seems to be reasonably consistent from species to species including man. Plasma protein binding tends to be less extensive for many drugs in the small mammalian species and somewhat more extensive in man.

The major discrepancy lies in the step that includes drug metabolism. Many common laboratory animal species, primarily the mouse and the rat, metabolize many lipid-soluble drugs or foreign compounds at an extremely rapid rate, whereas man metabolizes these drugs at a relatively slow rate.[1] The enzymes that metabolize drugs are relatively nonspecific with respect to the type of compounds they attack. In general, it seems very likely that once a drug reaches its site of action, its intrinsic activity is quite similar from species to species. The main problems in extrapolating results from one animal species to another arise in the step involved in drug metabolism. Drug metabolism in relation to species variation is a topic to which hours of discussion could be devoted. It should be pointed out, of course, that occasionally differences are noted in drug action related to the site of action. For example, we have studied the morphine-like immobilizing compound, etorphine (M99). A few milligrams of etorphine will tranquilize an elephant,[15] whereas doses approaching 1 milligram/kilo produce no detectable biological effects in sharks. We have reason to believe that this difference is not due to absorption, distribution, metabolism, or excretion, but probably is related to the relatively uncomplex central nervous system of the primitive elasmobranch as opposed to that of the much more highly evolved mammals.

The question to be asked, then, is "How well, in fact, do the animal models predict for the toxicity or action of drugs in man?" Let us first consider the quantitative aspect—i.e., compare the maximum tolerated doses for a variety of drugs in common laboratory animals and man. To explore this question, we studied such doses of anticancer drugs in man, monkey, and dog, and of LD_{10} in rat and mouse.[4] There are two important aspects to this comparison. First, anticancer drugs are used in man at dosage levels that cause tolerable toxicity. Thus we were able to estimate, with considerable precision, what the maximum tolerated dose is in man. Second, most of these antitumor agents are not involved in the variable drug-metabolizing enzymes because they tend to be lipid-insoluble and either act by interfering with intermediary metabolism or by reacting with biologically important macromolecules. The variability of hepatic microsomal metabolism was essentially eliminated by choice of this class of compounds. In a real sense this study tested variation as a function of size of animal and of absorption, distribution, and excretion. Using 15–20 compounds, the correlation coefficients ranged well above .9 for LD_{10} in laboratory animals versus the maximum tolerated dose in man with all 15–20 drugs representing threefold log range of doses. When these data are expressed as milligrams/kilo body weight, the regression line does not go through the origin. On the other hand, if the dosages are expressed in milligrams/sq meter (or as a function of a 2/3 power body weight), the location of the cluster points did move through the origin. A 12-fold factor between man and mouse relates surface area to body weight between these species. For one drug at least, methotrexate, it can be shown that the integrated plasma concentration time function after the same milligram/kilo dose is about 10 times higher in man than in the mouse. This probably is a function of what might be considered a scaling-up factor that is related to the size of the animal and to the relative velocity of the circulation as a function of animal size. These results underscore, of course, the need to study metabolism and plasma concentrations of drugs in laboratory animals and to try to relate them to what occurs in man.

The predictability of qualitative aspects of the biological effects, toxic or therapeutic, from laboratory animals to man, of anticancer drugs have also been studied. In a recent investigation [reported in abstract form, and being readied for publication by Philip Schein and co-workers* in my laboratory], it was found that predictability was quite good, indeed, for systems such as bone marrow, liver, kidney, and lung. When both dogs and monkeys were used, and if general toxicity to organ systems was considered rather than specific indices of toxicity, such as an increased bromsulphalein retention, for 20 drugs studied there was only one instance of failure to predict in all of these systems. This failure occurred with respect to renal toxicity. Many false positives occur, of course, in which the animal effect predicted organ-system damage that was not found in man. This difference may be accounted for by the fact that, in the laboratory animals, doses were always sufficient to cause the death of all of the animals; thus, every opportunity for exhibiting or demonstrating the full toxic potential of the compound was afforded.

In all 20 instances, problems of the central nervous system were predicted if one considered the organ system in general. With one compound, convulsions were observed at very high doses in the laboratory animals, whereas sleeping and impending coma were observed at moderately high doses in man.

It would seem, therefore, that in general the laboratory animals predict well for the effects of drugs in man. But they, like all biological systems, are not perfect and any pharmacologist, toxicologist, or clinician who attempts to extrapolate information on drug action from laboratory animals to man must be exceedingly cautious.

---

*Schein, P., Davis, R., Carter, S., Newman, J., and Rall, D. 1970. The toxicologic evaluation of anticancer drugs in dogs and monkeys as a basis for the prediction of qualitative toxicities in man. Clin. Pharmacol. Exp. Ther. 11:3–40.

PROBLEMS OF ANIMAL SYSTEMS

Next, I shall discuss some of our problems in developing an animal model useful for the discovery and evaluation of antitumor agents prior to their use in man.

After many years, an animal model was developed that provided an efficient screening system and secondary evaluation system for studying compounds of potential antitumor activity. This is the L1210 lymphoid leukemia, which grows in mice.[5] Recently I reviewed what might be called the successes and failures of cancer chemotherapy in man and it is apparent that the successes have largely come in human tumors that proliferate relatively rapidly.[16] It has recently been recognized that most tumors are not composed entirely of rapidly dividing cells. In fact, the majority of large solid carcinomas in man act as if they are composed of a relatively small fraction of cells that are dividing at a comparatively rapid rate, with perhaps 1–5 percent of the cells having a doubling time of a few days. Many of the viable cells are not in a proliferative state at all. The L1210 leukemia is completely different and is a very rapidly proliferating tumor with 90–95 percent of the cells in cell cycle.

Clinical diseases, in which therapeutic success ranging from cure to long-term remissions has occurred, also involve those tumors in man that multiply at the greatest rate, and have a very high fraction of proliferating cells. It seems apparent that the kinetics of the tumor growth may well be more important than the morphologic character of the tumor. It took many years to establish this relationship and we are now searching for, and establishing, a series of slowly proliferating tumors in animals that we hope may prove to be better models for the more common slowly proliferating tumors in man.

SUMMARY

The use of laboratory animal models for the study of drug therapy requires, first, an appreciation of the problems of drug effects in laboratory animals and in man. These effects can best be considered in terms of similarities or differences with respect to absorption, distribution, excretion, metabolism, and arrival at and effect on ultimate site of action. Of these factors, drug metabolism is frequently disparate between laboratory animals and man. Evidence has been obtained, through a careful comparison of doses and effects, that between laboratory animals and man in general there is good agreement.

In the search for effective anticancer drugs, it is becoming apparent that rate of cellular proliferation may be the most important factor in determining responsiveness to drugs. Present transplanted tumor screens predict well for rapidly proliferating human tumors, because we believe these model tumors are proliferating rapidly. Encouraging clinical successes in rapidly proliferating tumors, using drugs developed from these screens, suggest that this factor may be more important than morphological or biological type or classification. Thus, current studies are aimed at developing new laboratory animal tumor models that correspond to the slowly proliferating common human neoplasms.

REFERENCES

1. Brodie, B. B. 1964. Of mice, microsomes and men. Pharmacologist 6:12–18.
2. Burgen, A. S. V. 1965. The predictive value of animal toxicity tests. Proc. Second Internat. Pharmacol. Meeting 8:49–56.
3. DeSomer, P., H. Van De Voorde, H. Eyssen, and P. Van Dijck. 1955. A study on penicillin toxicity to guinea pigs. Antibiotics and Chemotherapy 5:463–469.
4. Freireich, E. J., E. A. Gehan, D. P. Rall, L. H. Schmidt, and H. E. Skipper. 1966. Quantitative comparison of toxicity of anticancer agents in mouse, rat, hamster, dog, monkey, and man. Cancer Chemotherap. Rep. 50:219–244.
5. Goldin, A., A. A. Serpick, and N. Mantel. 1966. Experimental screening procedures and clinical predictability value. Cancer Chemotherap. Rep. 50:173–218.
6. Haladay, R. 1958. The absorption of antibodies from immune stems from the gut of the young rat. Proc. Royal Soc. (Biol.) 143:408.
7. Hamre, D. M., G. Rake, C. M. McKee, and H. B. MacPhillamy. 1943. The toxicity of penicillin prepared for clinical use. Amer. J. Med. Sci. 206:642–652.
8. Heilman, F. R. and W. E. Herrell. 1944. Penicillin in the treatment of experimental leptospirosis icterohemorrhagica. Proc. Mayo Clinic 19:89–99.
9. Henderson, E. S., R. H. Adamson, C. Denham, and V. T. Oliverio. 1965. The metabolic fate of methotrexate-H[3]. I. Absorption, excretion and distribution in mice, rats, dogs and monkeys. Cancer Res. 25:1008–1017.
10. Henderson, E. S., R. H. Adamson, and V. T. Oliverio. 1965. The metabolic fate of methotrexate-H[3]. II. Absorption and excretion in man. Cancer Res. 25:1018–1024.
11. Hogben, C. A. M., L. S. Schanker, D. J. Tocco, and B. B. Brodie. 1957. Absorption of drugs from the stomach. II. The human. J. Pharmacol. Exp. Ther. 120:540.
12. Hume, A. S., M. T. Bush, J. Remick, and B. H. Douglas. 1968. Comparison of gastric absorption of thiopental and pentobarbital in rat, dog, and man. Arch. Int. Pharmacodyn. Therap. 171:122–127.
13. Koppanyi, T. and M. A. Avery. 1966. Species differences and the clinical trial of new drugs. A review. Clin. Pharmacol. Ther. 7:250–270.

14. Marshall, E. K., Jr. 1930. A comparison of functions of the glomerular and aglomerular kidney. Amer. J. Physiol. 94:1.
15. Rall, D. P. 1967. Comparative pharmacology and cerebrospinal fluid. Proc. International Symposium on Comparative Pharmacology. January 24–27, 1967, Washington, D.C. Fed. Proc. 26:1020–1023.
16. Rall, D. P. New approaches in administration of anticancer drugs. Cancer Res., in press.
17. Rall, D. P. and J. W. Burger. 1967. Some aspects of hepatic and renal excretion in myxine. Amer. J. Physiol. 212:354–356.
18. Schanker, L. S., P. A. Shore, B. B. Brodie, and C. A. M. Hogben. 1967. Absorption of drugs from the stomach. I. The rat. J. Pharmacol. Exp. Ther. 120:528.
19. Welch, A. D. 1967. Pharmacological differences, qualitative and quantitative, between man and other species. In: *Drug Responses in Man,* ed. by G. Wolstenholme and R. Porter, pp. 3–23, Little, Brown, Boston.

The Use of Marine Mammals in Biomedical Research

RICHARD C. HUBBARD

INTRODUCTION

The 117 species of marine mammals are divided into four orders: whales, seals, sea otters, and sirenians (dugongs and manatees).[35] We will consider here only seals (or pinnipeds) and whales. The pinnipeds include three subdivisions: true seals, eared seals (sea lions and fur seals), and walrus. Whales (or cetaceans) are the fluked mammals, which are divided into two major categories—baleen and toothed whales. All baleen whales are very large; only the smaller toothed whales (porpoises, dolphins, and animals up to the size of the pilot and killer whale) are held in captivity.

Before we can determine how these animals can be used in biomedical research, we need to understand their characteristics. Marine mammals, which at first glance seem so very different from terrestrial mammals, have in fact characteristics that are adaptive modifications of the characteristics of the latter. These modifications frequently magnify processes, making them easier to study. To illustrate, initially there was no indication whatsoever that the remarkable diving ability of marine mammals actually depends on a magnified oxygen-conserving mechanism that is possessed to lesser degrees by man and most other vertebrates.[37]

THE DIVING REFLEX

Since 1838, when Burow[6] described the seal's abdominal blood reservoir and its dam, the posterior vena caval valve (so well illustrated by

Harrison and Tomlinson,[11] scientists have sought to explain why marine mammals are able to remain submerged longer than man. Increased oxygen storage is not the answer.[7, 36] Taking into consideration all of the oxygen stores within the body, it has been shown[10] that seals can muster, on a pound-for-pound basis, perhaps twice the reserves available to man; yet they are capable of remaining under water for 20 to 40 minutes. Whales have been reported to submerge for much longer periods.

Only in the last 30 years or so have the mechanics of this diving reflex phenomenon been explained.[2, 15] It is in essence a circulatory adjustment that conserves oxygen. When an animal dives, a generalized arterial constriction occurs outside the brain-heart-lung cycle, i.e., in areas least sensitive to lack of oxygen. The animal exhibits what Van Slyke refers to as a "heart-lung-brain preparation."[37]

The tissues subjected to vascular constriction are prevented from gaining access to blood oxygen. Consequently, they operate in an oxygen-debt situation, conserving the blood oxygen for the brain and the heart, the two tissues most sensitive to anoxia. As a result of this generalized vasoconstriction, there is a marked decrease of the "functional" vascular space, and the heart must slow down in order to keep blood pressure within normal limits. This diving bradycardia is the first sign of the diving reflex that early physiologists noted.[4, 32] It is so marked in the true seal that the rate often is as low as, or lower than, one-tenth of the nondiving rate; yet the blood pressure remains unchanged or is even slightly elevated. The reduced rate suggests that as much as 90 percent of the body tissues are affected by the arterial vasconstriction.

The tissues subjected to vasoconstriction build up lactic acid and other anaerobic metabolites. At the end of the dive, when the animal commences to breathe, the mass constriction is released, the blood is flooded with these metabolites, and the blood pH drops. The fact that the seal is again prepared to dive after taking only a few breaths of air suggests that very efficient mechanisms operate to control acidosis and the buildup of CO_2.

The diving reflex is not confined to marine mammals. Man is also endowed with this ability, though to a much lesser degree.[2] Yogis who remain sealed in coffins under water for several hours develop a marked bradycardia.[1] The heart rates of native Australian skin divers have been reported to drop 50 percent, though dives were usually less than 1 minute in duration.[38] In man, however, the diving reflex is accomplished by arrhythmias and altered ECG.[2]

THE RELATIONSHIP OF THE DIVING REFLEX TO DEATH IN MAN

An exaggerated diving reflex response may help to explain some types of death in man. Wolf[47] has shown that many diseases of man are simply reflections of body reactions to the environment. Sometimes these reactions involve inappropriate responses of usually effective defense mechanisms. There have been cases of sudden death from supposed heart attack where no significant coronary lesions or pathology of any nature could be found. Wolf suggests that in such cases death could have resulted from an exaggerated "diving reflex" response. He also suggests this exaggerated response may be a contributory mechanism of death in cases of fear and social ostracism (voodoo death).[48, 49]

MODELS FOR SHOCK THERAPY STUDIES

The circulatory changes experienced by man in shock are nearly identical to those experienced by a diving seal or whale. They include massive vascular constriction, central venous pooling, the confinement of circulation primarily to the brain-heart-lung cycle, blood sludging, lowered blood pH, and electrolyte alterations; the most disastrous of these is the massive vascular constriction. Fine[8] describes the central lesion of shock as being unrelenting vascular constriction. Why this constriction should persist in irreversible shock is one of the great mysteries of that condition, and little advance has been made toward solving the problem. In the seal the mechanism turns on and off each time it dives and surfaces. If we can understand the neural physiology and the biochemistry operative in the seal, it seems reasonable to suppose that we will better understand the unrelenting constriction of shock.

Wendt *et al.*[46] speak of the importance of body biochemistry (metabolic factors) in cardiac output, and of the fate of ADP and ATP in anoxia. The biochemistry of the seal while it is defending itself against anoxia has scarcely been studied. It is a promising field, with important implications for the management of shock. How is it that an animal can utilize its muscles under anoxic conditions for extended periods, and yet, after a brief time, be again prepared to dive? Part of the explanation may lie in modified enzyme systems (for example, modifications in the ADP-ATP system). There is considerable evidence that these animals possess modified biochemical processes in a variety of body functions. Robin *et al.*[33] and Murdaugh *et al.*[18] have expressed an interest in their glycolytic pathways. Undoubtedly, some of these modifications play an important role in diving ability.

Much time and effort have been expended to develop buffers for acidotic shock patients. An understanding of the ability of marine mammals quickly to divest themselves of anaerobic metabolites may direct us to better methods of handling such human patients.

In the past, central venous pooling has been considered as one of the central lesions in shock. However, *if the seal is indeed a model,* it seems just as likely that this phenomenon simply accommodates for the vast peripheral constriction that is taking place. The seal has anatomical features to accommodate this central pooling in the form of its voluminous posterior vena cava and unique hepatic sinus (see photograph, p. 241, Ref. 10), along with numerous other exceptionally flaccid veins. These reservoirs apparently cannot contract. It seems reasonable that Burow's posterior vena caval valve may close during diving, preventing this reservoir blood from overflooding the slowed heart. If this is so, then central pooling is a normal consequence of vascular constriction and ebbs and flows with the presence or absence of peripheral constriction.

If the persistent constriction of shock can be released, it would seem that this should simultaneously correct central pooling, blood sludging, and lack of oxygen at the cellular level. Although some hesitate to compare the circulation changes in diving and in shock,[12] a belief that some relationship does in fact exist seems to me inescapable. It thus appears very pertinent to study the mechanisms controlling vascular constriction in seals for clues to ways of reducing the unrelenting constriction of shock.

COAGULATION

An understanding of blood coagulation characteristics of marine mammals may be useful in the development of drugs to reduce clotting time at the site of surgical hemorrhage, and even to aid patients with blood clotting deficiencies.

To recognize the potentials of such a study, we again need to define the characteristics of the marine mammal. It has been noted that clotting times of different blood samples vary greatly, even from the same species of animal.[12, 34] In some cases it was not uncommon to have the blood coagulate so rapidly that it could not be mixed with anticoagulant. In other circumstances, blood might take two or more hours to coagulate.[30] Some order appeared in these findings when we were able to compare the clotting time of blood drawn directly from a vessel ("clean" blood) with the clotting time of blood collected from a wound or through a needle "contaminated' with extravascular humors. "Clean" blood coagulated slowly in comparison with "contaminated" blood. In later work, Dr. Judith Pool of Stanford Medical Center studied five blood-plasma clotting factors from three elephant seals and found no indication of any unusual concentrations of these factors. She also reported that Jessica Lewis, in more extensive but also unpublished work with the harbor seal, made similar findings. Still later, Dr. Jean Robinson of San Francisco General Hospital discovered that cetacean blood plasma lacks Hageman Factor. It now seems that an extrinsic (tissue) clotting factor may be responsible for the exceptionally rapid clotting seen in wounds and first-drawn I.V. samples.

We also believe we understand why marine mammals have such a "double" mechanism. During diving some sludging of blood flow probably occurs (see photo, page 241, Ref. 10), and there must therefore be protection from intravascular clotting. It appears that marine mammals have so adjusted their intrinsic clotting factors (those located in the bloodstream itself) that the danger of intravascular clotting from sludging is eliminated. On the other hand, such free-flowing blood would be intolerable to a wounded animal in the water. We believe that the extrinsic blood clotting factors (those outside the blood vessels) compensate for any "deficiency" of vascular constituents in a highly efficient manner so that, once intrinsic and extrinsic factors are combined (as in a wound), clotting is exceptionally rapid. This theory is in accord with our experience; the first-drawn blood and capillary blood (which were most apt to be "contaminated" with extravascular humors) invariably clotted rapidly, while blood drawn after the initial puncture did not clot quickly.

To exploit these potentials, one should first analyze the differences between the intrinsic and extrinsic factors of marine mammals, and then compare the findings with those for terrestrial mammals. It may then be possible to ascertain which factor or mechanism is responsible for the rapid clotting in marine mammals. We will be well rewarded if such information leads to the development of pharmaceutical products that aid surgeons in controlling troublesome hemorrhage, or possibly aid patients whose blood does not clot quickly enough. A comparative study of the coagulation factors of marine mammals and terrestrial mammals would also give a broader perspective of comparative biology.

PHARMACOLOGY

Of interest to the pharmacologist and the physiologist is the fact that the pharmacodynamics of marine mammals is sometimes peculiar. The subject is covered in greater detail elsewhere,[14] but a few examples will illustrate the potential of seals and whales as biomedical models.

To date, it has been impossible to completely dilate the eye of any seal with topical mydriatics. Atropine has been used at hourly intervals for five applications. Cyclopentolate HCl (Cyclogolyl, Scheffelin), homatropine 5 percent, and tropicamide 1 percent (Mydriacyl, Alcon) have also been employed. On the other hand, oral administration of a single dose of methscopolamine has paralyzed the pupil for as long as 30 days.

Exceptionally small doses of tranquilizer produce effects in whales. Phenothiazine-derived tranquilizers are toxic to seals.

Neither apomorphine HCl nor emetine HCl will produce emesis in seals or whales.

An understanding of these and other drug-related peculiarities of the marine mammals should have important implications for the advancement of knowledge in the field of human pharmacology.

The pharmacological uniqueness of marine mammals is surely a reflection of biochemical and neurological modifications adapting them to an aquatic life. We view these changes in drug reaction not as completely different from man, but rather as modifications of processes also present in man. If such is the case, then an elucidation of the details of these modifications will lead us to a better understanding of the drug actions in man. A dramatic result of such biochemical and pharmaceutical studies would be the development of new therapeutic techniques to combat shock and other human disorders.

MODELS FOR FAT METABOLISM STUDIES

There are several indications that the fat metabolism and fat utilization processes of marine mammals are different from those of terrestrial mammals. A greater part of marine mammals' energy may come directly from fat utilization.

Among the pinnipeds, the harem bull is a case in point. This active animal must defend his territory and harem against all comers, yet he fasts for several weeks during this period. His only obvious source of energy is stored fat, which visibly diminishes as the breeding season progresses. Only in hibernating animals do we see anything comparable, but the slow conversion in the latter case is not really parallel to that of the active harem bull. Because he does not undergo a reduction in metabolic rate, the bull must convert fat very rapidly to meet energy requirements for defending his harem and for breeding.

A similar situation exists with the nursing cow, which remains with her newborn pup for several days without eating. It is probable that the high energy requirements during this period are met almost exclusively by conversion of fat.

Puppione[29, 30] has shown that the serum lipoprotein distribution in pinnipeds is markedly different from that reported for other animals, not only in terms of serum concentration in certain density ranges but also in terms of percent composition of the lipid moiety. He has further shown that the enzyme lecithin cholesterol acyl transferase (LCAT), associated with cholesterol ester metabolism, is much lower in these animals than in man.

Fat deposits in marine mammals also are different from those in terrestrial mammals. For example, there are no fat deposits around the heart or kidneys or in the mesenteries of marine mammals.[44]

The milk of all marine mammals is high in fat. It seems particularly significant that the milk of the eared seals (sea lions) is devoid of milk sugar,[17] yet the blood glucose levels of nursing pups must be maintained at even higher levels than those for man.[13] Experience has shown that blood glucose levels below 65 mg percent may produce symptoms of hypoglycemia in young pups.[12]

We would conventionally assume that the high levels of blood glucose in the pups derive from the milk protein; Sinclair[41] has suggested, however, that due to their very high intake of polyunsaturated fat, Eskimos may be converting these polyunsaturates to pyruvic acid, which could give a net glucose gain. Perhaps the seal, whose diet is high in polyunsaturated fats,[30] would be an excellent model in which to study such a pathway.

If a modification of fat utilization and storage could be demonstrated, it might shed light on such problems of human health as endogenous obesity and the relationship of blood cholesterols to atherosclerosis.

MODEL FOR THE STUDY OF EYE DISEASES

Blindness is not incompatible with survival in seals and sea lions, as it is in most terrestrial animals. Well-nourished blind pinnipeds are observed in the wild, and it is presumed that they compensate by using sonar. This eliminates starvation—the usual consequence of blindness in terrestrial animals—as a means of natural selection against hereditary eye weaknesses. The frequency of bilateral eye problems in wild pinnipeds suggests that a natural predisposition to eye disease does in fact exist. Benchley[3] reports that a "milky" eye is common in marine mammals. Bonnet,[5] Florkin and Redfield,[9] and Ward[45] also refer to blind pinnipeds in the wild. Lockley (cit. Pearson[21] p. 55) has found totally blind seals in excellent health, and Harrison (personal communication) has observed the same. Pearson[21] contends that the harbor seal has excellent vision out of water but doubts if it is able to use this sense in the muddy waters of the Wash (England). Poulter[22, 23] was first led to consider the use of sonar by sea lions by observing blind but well-nourished animals on Año Nuevo Island.[20] Bilateral eye conditions observed by the author in wild pinnipeds on Año Nuevo Island were squinting and photophobia, corneal opacities, pigmentation (or depigmentation) of the iris, cataracts, and neural degeneration characterized by completely dilated pupils, absence of squinting, and a bright, easily observed tapetum lucidum.

One of the commonest problems of captive pinnipeds is eye disease.[13] The pinniped eye is a museum of diseases ranging from conjunctivitis, keratitis, and luxated lenses to unidentified iris changes, cataracts, and neural degeneration. We have found therapy difficult and satisfactory dilation of the pupil nearly impossible.[14]

In rearing Steller pups at Stanford Research Institute, two of four were found to have rapidly maturing bilateral cataracts at the age of seven months. Examination of the remaining two animals' lenses revealed various degrees of bilateral lens opacity. Several circumstances have led us to consider the possibility that here we have a model for the study of glucose cataracts. The formula fed to these pups contained both lactose from the added cow's cream, which is not generally regarded as digestible by otarids, and added amounts of glucose to combat hypoglycemia.[12] Immediately upon discovering the cataracts, the use of cream and the added glucose in the food formula was discontinued. The cataracts ceased further development, and nearly a year later the previously immature cataracts of the last two animals had changed little, if any. In demonstrating sugar cataracts in rats, two characteristics are typical; first, only young animals are susceptible, and second, the sugar cataracts develop rapidly.[16] With the Steller sea lions too, the opacities appeared very rapidly, and in young animals. It must be acknowledged that occasionally wild animals that have not been subjected to artificial diets have also demonstrated bilateral mature cataracts. But if a hereditary predisposition to cataracts does exist, and if excess glucose can be shown to excite cataract production, then we may have a model that closely replicates the diabetic cataract.

SONAR

Humans have sonar or echo-locating ability.[31] It has long been recognized that certain blind individuals are exceptionally mobile. For example, one blind teenager rides his bicycle on the street and can also locate and tackle a person carrying a football, and a blind man works as a tree-topper. These people operate by what has been referred to as "facial vision." But this facial vision disappears if the ears are plugged. The fact is that this ability involves the use of sonar, or echo location. Each of us has this capability to some degree. If one closes his eyes, alternatively holds a disk of metal and a stretched hoop of paper in front of his face, and makes a sound, such as hissing or clicking, he can learn to distinguish one object from the other by the difference in the returning echo. However, only a small number of blind humans have learned to use this ability efficiently.

Seals and whales possess an amazingly sophisticated sonar capability. Figure 1 shows a blindfolded dolphin swimming through an obstacle course without touching the obstacles.[19] Sea lions can also avoid obstacles under similar conditions, and their sonar has been shown to be so acute that they can distinguish between two species of fish, one of which they prefer over the other.[24, 28] They possess a natural echo-ranging ability far superior to anything man has yet artificially created. If we can learn from these biological models how to interpret the echo information, we may be able to develop a portable electronic system to help the blind "navigate."

Scientists have demonstrated the outstanding sonar ability of marine mammals in many ways. The most recent advance has been the application of sophisticated physical techniques to the analysis of the primary sonar signal, especially in the case of the California sea lion.[24, 26–28, 39, 40, 42] Future progress requires that we analyze the returning echoes of these primary signals. To this end there has been constructed at Stanford Research Institute (see Figure 2) the most advanced anechoic research tank in the nation; it absorbs frequencies down to 300 cycles per second and has a sound decay rate of 4,500 decibels per second.[25] Correlation of the physical analysis of primary sonar signals with that of their echoes may well tell us many of the requirements for a successful artificial sonar system for the blind. There is every reason to believe that future echo-analysis research will be as fruitful as past analysis of primary sonar signals.

FIGURE 1   Blindfolded dolphins repeatedly negotiate mazes without contacting obstacles. (Photo: Richard Hewitt.)

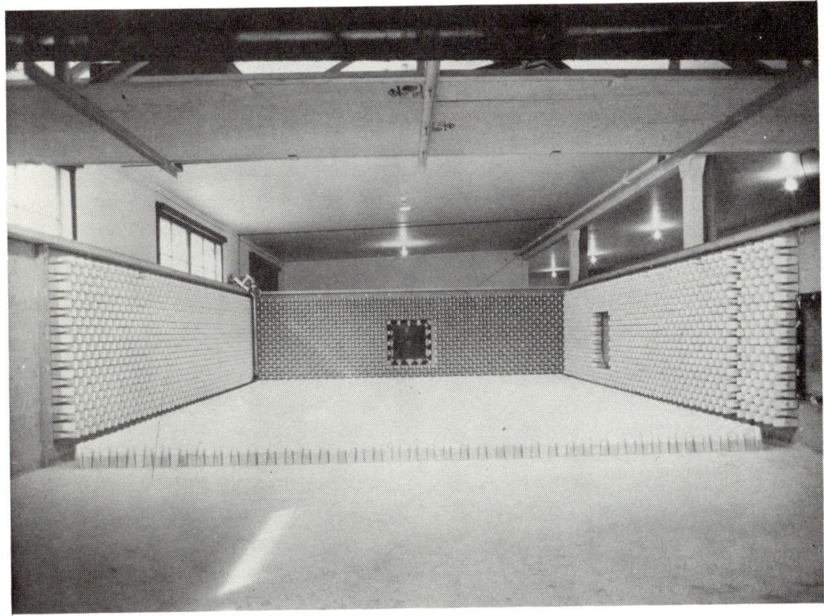

FIGURE 2   View of one end of empty 17 ft by 80 ft anechoic tank showing distribution of 6-inch long wedges that are fastened to the walls and bottom for a distance of 20 ft at each end of the tank. Sand inside 6-inch walls absorbs sound energy.

ANATOMY

The heart of a whale closely resembles a very large human heart. A 200-pound whale heart allows the study of details and relationships applicable to human heart problems and disease that cannot be examined in the small human heart. For example, a good differentiation of conduction fibers in the whale has facilitated a comparative study of each of the elements of the conduction system (i.e., SA and AV nodes, AV bundle, and bundle branches).[43]

CONCLUSION

The studies and observations reported in this brief review support the position that there is a very useful future for marine mammals as biomedical models. The technological skills needed to define these potentials more clearly are available, and will without question be applied more and more by a growing segment of the scientific community.

ACKNOWLEDGMENT

I am deeply indebted to Thomas C. Poulter for affording me the opportunity to make broad spectrum studies of marine mammals. I appreciate the permission of Judith Pool and Jean Robinson to report on their unpublished coagulation studies. Donald Puppione has been an invaluable critic in aspects related to his field of study. And last but not least the patience of my associates, Robert Pollard and Lawrence Martinez, and my secretary, Patricia Howard, should not go unnoticed.

This investigation was supported in part by Public Health Services Research Grant NB 04736 from the Division of Research Grants, National Institutes of Health.

REFERENCES

1. Anand, B. K., G. S. Chhina, and B. Singh. 1961. Studies on Shri Ramanand yogi during his stay in an air-tight box. Indian J. Med. Res. 49:82.
2. Andersen, H. T. 1966. Physiological adaptations in diving vertebrates. Physiol. Revs. 46:212–243.
3. Benchley, F. J. 1930. Experience with elephant seal. Parks and Recreation 13(5):317–320.
4. Bert, Paul. 1879. Leçons sur la physiologie comparée de la respiration. Baillière, Paris.
5. Bonnot, P. 1928. The sea lions of California. California Fish and Game 14:1–17.
6. Burow. 1838. Das Gefässystem der Robben. Archiv. für Anatomic, Physical Wassenschaff-liche Medicine: 230–258.
7. Cross, C. E., B. S. Packer, M. A. Altman, *et al.* 1967. Determination of total body oxygen stores in normal man and seals. J. Clin. Invest. 46:1048.
8. Fine, J. 1965. Current status of the problem of traumatic shock. In *Shock and Hypotension: Pathogenesis and Treatment,* L. C. Mills and J. H. Moyers, eds., pp. 1–9, Grune and Stratton, New York.
9. Florkin, M. and A. C. Redfield. 1931. On the respiration function of the blood of the sea lion. Biol. Bull. 61:422–426.
10. Harrison, R. J. and G. L. Kooyman. 1968. General physiology of pinnipedia. In *Behavior and Physiology of Pinnipeds,* Eds., R. J. Harrison, R. C. Hubbard, R. S. Peterson, Charles E. Rice, and R. J. Schusterman. Appleton-Century-Crofts, New York.
11. Harrison, R. J. and J. D. Tomlinson. 1956. Observations on the venous system in certain pinnipedia and cetacea. Proc. Zool. Soc. London 126, part 2:205–233.

12. Hubbard, Richard C. and Thomas C. Poulter. 1968. Seals and sea lions as Models for Studies in Comparative Biology. Laboratory Animal Care 18, Part II:288–297.

13. Hubbard, Richard C. 1968. Husbandry and Laboratory Care of Pinnipeds. In *Behavior and Physiology of Pinnipeds,* Eds., R. J. Harrison, R. C. Hubbard, R. S. Peterson, Charles E. Rice, and R. J. Schusterman. Appleton-Century-Crofts, New York.

14. Hubbard, Richard C. 1969. Chemotherapy in Marine Mammals. Wildlife Disease (in press).

15. Irving, L. 1939. Respiration in diving mammals. Physiol. Rev. 19:112–134.

16. Lerman, S. 1965. Metabolic pathways in experimental sugar and radiation cataract. Physiol. Rev. 45:98–117.

17. Mathai, C. K., M. E. Q. Pilson and E. Beutler. 1966. Galactose Metabolism in the Sea Lion. Proc. Soc. of Exp. Biol. and Med. 123:603–604.

18. Murdaugh, H. V., Jr., E. D. Robin, J. E. Miller, W. F. Drewry, and E. Weiss. 1966. Adaptations to diving in the harbor seal: cardiac output during diving. Am. J. Physiol. 210:176–180.

19. Norris, Kenneth S., John H. Prescott, Paul V. Asa-Dorian, and Paul Perkins. 1961. An Experimental Demonstration of Echo-Location Behavior in the Porpoise, Tursiops truncatus (Montagu). The Biological Bulletin 120:163–176.

20. Orr, Robert T. and Thomas C. Poulter. 1962. Año Nuevo Marine Biological Park. Pacific Discovery XV(1):13–19.

21. Pearson, R. H. 1961. *A Seal Flies By.* Walker and Company, New York.

22. Poulter, Thomas C. 1963. Sonar Signals of the Sea Lion. Science 139:753–755.

23. Poulter, Thomas C. 1963. The Sonar of the Sea Lion. Transactions in Ultrasonic Engineering IEEE-UE-10:109–111.

24. Poulter, Thomas C. 1966. Use of Active Sonar by the California Sea Lion. Journal of Auditory Research 6:155–173.

25. Poulter, Thomas C. 1968. Anechoic Biological Sonar Research Tank. Section XVII, 1–18 in Progress Report 1968, Echo Ranging Signals, SRI Project No. 6830, NIK Contract NB 04736-05.

26. Poulter, Thomas C. 1968. Underwater Vocalization and Behavior of Pinnipeds. In *The Behavior and Physiology of Pinnipeds.* Eds., R. J. Harrison, R. C. Hubbard, R. S. Peterson, Charles E. Rice, and R. J. Schusterman. Appleton-Century-Crofts, New York.

27. Poulter, Thomas C. 1969. Sonar of Penguins and Fur Seals. Proceedings of the California Academy of Sciences, Fourth Series XXXVI: 13, 363–380.

28. Poulter, Thomas C. and R. A. Jennings. 1969. Sonar Discrimination Ability of the California Sea Lion, Zalophus Californianus. Proceedings of the California Academy of Sciences, Fourth Series XXXVI: 14, pp. 381–389.

29. Puppione, D. L. and E. L. Coggiola. 1966. Serum lipoprotein distribution in certain aquatic mammals. In Proceedings of the Third Annual Conference on Biological Sonar and Diving Mammals, C. E. Rice, Ed., pp. 34–60, Stanford Research Institute, Menlo Park, California.

30. Puppione, D. L. 1969. Physical and Chemical Characterizations of the Serum Lipoproteins of Marine Mammals. Lawrence Radiation Laboratory. University of California, Berkeley, UCRL 18821.

31. Rice, Charles E. 1967. Human Echo Perception. Science 155(3763):656–664.
32. Richet, Charles, 1894a. La Resistance des canards a l'asphyxie. C. R. Soc. Biol. Paris 1:244–245.
33. Robin, E. D., J. V. Murdaugh, Jr., W. Pyron, E. Weiss, and P. Soters. 1963. Adaptations to diving in the harbor seal—gas exchange and ventilatory response to CO_2. Am. J. Physiol. 205:1175–1177.
34. Scheffer, V. B. 1963. *Seals, Sea Lions and Walruses.* Stanford University Press, California.
35. Scheffer, Victor B. and Dale W. Rice. 1963. A List of the Marine Mammals of the World. United States Fish and Wildlife Service, Special Scientific Report, Fisheries No. 431, Washington, D.C.
36. Scholander, P. F. 1940. Experimental Investigations on the Respiratory Function in Diving Mammals and Birds. Hvalraadets-Skrifter, 22, pp. 7–131, Det Norske Videnskaps Akademi Oslo.
37. Scholander, P. F. 1963. The master switch of life. Scientific American. 209(6): 92–106.
38. Scholander, P. F., H. T. Hammel, H. Le Messurier, E. Hemmingsen, and W. Garey. 1962. Circulatory Adjustments in Pearl Divers. J. Appl. Physiol. 17(2):184–190.
39. Shaver, H. N. and T. C. Poulter. 1967. Sea lion echo ranging. Acoust. Soc. Am. 42(2):428–437.
40. Shaver, Harry N., Bruce M. Sifford, Jim K. Omura, N. Tom Gaarder, and Russell T. Wolfram. 1968. Analysis of Marine Mammal Signals. Section VII, 1–83 in Progress Report 1968, Echo Ranging Signals, SRI Project No. 6830, NIH Contract NB 04736-05.
41. Sinclair, H. M. 1964. Carbohydrates and Fats. In *Nutrition,* Vol. I, G. H. Beaton and E. W. McHenry, Eds., Academic Press.
42. Singleton, Richard C. and Thomas C. Poulter. 1967. Spectral Analysis of the Cell of the Male Killer Whale. IEEE Trans. On Audio and Electro Acoustics AU-15:2, 104–113.
43. Truex, Raymond C. and Martha Q. Smythe. 1965. Comparative Morphology of the Cardiac Conduction Tissue in Animals. Annals of the New York Academy of Sciences 127:19–33.
44. Vague, J. and R. Fenasse. 1965. Comparative Analogy of Adipose Tissue. Chapter 5 of *Adipose Tissue,* Section 5 of *Handbook of Physiology*. Albert E. Reynolds and George F. Cahill, Jr., Eds. American Physiological Society, Washington, D.C.
45. Ward, H. L. 1887. Notes on the life history of Monachus tropicalis, the West Indian seal. American Naturalist 21:257–264.
46. Wendt, V. E., C. Wu, R. Balcon, and R. J. Bing. 1965. Metabolic factors in the control of cardiac output. In *Shock and Hypotension: Pathogenesis and Treatment.* L. C. Mills and J. H. Moyer, Eds., pp. 32–38. Grune and Stratton, New York.
47. Wolf, Stewart. 1961. Disease as a Way of Life: Neural Integration in Systemic Pathology. Perspectives in Biology and Medicine 4:288–305.
48. Wolf, Stewart. 1966. Sudden Death and the Oxygen-Conserving Reflex. Am. Heart J. 71:840–841.
49. Wolf, Stewart. 1967. The End of the Rope: The Role of the Brain in Cardiac Death. Canad. Med. Ass. J. 97:1022–5.

Symposium Contributors

ROBERT T. ADER, Department of Psychiatry
University of Rochester School of Medicine and Dentistry, Rochester, New York

BILL C. BULLOCK, Department of Laboratory Animal Medicine
The Bowman Gray School of Medicine, Winston-Salem, North Carolina

THOMAS B. CLARKSON, Professor and Head
Department of Laboratory Animal Medicine
The Bowman Gray School of Medicine, Winston-Salem, North Carolina

CHARLES E. CORNELIUS, Dean, College of Veterinary Medicine
Kansas State University, Manhattan, Kansas

WILLIAM C. DAVIS, Department of Microbiology
College of Veterinary Medicine, Washington State University
Pullman, Washington

CAROLINE W. EASLEY, Department of Biochemistry, College of Medicine
J. Hillis Miller Health Center, University of Florida, Gainesville, Florida

JOHN H. GORHAM, Animal Disease and Parasite Research Division
Agriculture Research Service, U.S. Department of Agriculture
Pullman, Washington

GERALD A. HEGREBERG, Department of Veterinary Pathology
Washington State University, Pullman, Washington

JAMES M. HOLLAND, Department of Veterinary Pathology
College of Veterinary Medicine, Washington State University
Pullman, Washington

RICHARD C. HUBBARD, Stanford Research Institute
333 Ravenswood Avenue, Menlo Park, California

HYRAM KITCHEN, Division of Comparative Medical Research
Center for Laboratory Animal Resources, A-46 Veterinary Clinic
Michigan State University, East Lansing, Michigan

ETHEL OLIVER KOUBA, Department of Biochemistry, College of Medicine
J. Hillis Miller Health Center, University of Florida, Gainesville, Florida

NOEL D. M. LEHNER, Assistant Professor
Department of Laboratory Animal Medicine
The Bowman Gray School of Medicine, Winston-Salem, North Carolina

HUGH B. LOFLAND, Professor of Pathology (Biochemistry)
The Bowman Gray School of Medicine, Winston-Salem, North Carolina

JOHN E. LUND, Department of Pathology and Animal Care Facility
Stanford Medical School, Stanford, California

MERLE E. MUHRER, Department of Agricultural Chemistry
University of Missouri, Columbia, Missouri

GEORGE W. NACE, Department of Zoology &
The Center for Human Growth & Development, The University of Michigan
Ann Arbor, Michigan

GEORGE A. PADGETT, Department of Veterinary Pathology
College of Veterinary Medicine, Washington State University
Pullman, Washington

ROY C. PAGE, Department of Pathology, School of Medicine
University of Washington, Seattle, Washington

ROBERT W. PRICHARD, Professor, Department of Pathology
The Bowman Gray School of Medicine, Winston-Salem, North Carolina

DAVID J. PRIEUR, Department of Veterinary Pathology
College of Veterinary Medicine, Washington State University
Pullman, Washington

DAVID P. RALL, Associate Scientific Director for Experimental Therapeutics
National Cancer Institute, National Institutes of Health
Bethesda, Maryland

RICHARD W. ST. CLAIR, Assistant Professor of Pathology (Physiology)
The Bowman Gray School of Medicine, Winston-Salem, North Carolina